To Cind...

Joan Easton Lentz

Ellen Easton

CALIFORNIA NATURAL HISTORY GUIDES

**INTRODUCTION TO
BIRDS OF THE
SOUTHERN CALIFORNIA COAST**

California Natural History Guides

Phyllis M. Faber and Bruce M. Pavlik, General Editors

Introduction to
BIRDS of
the Southern
California Coast

Joan Easton Lentz

Don DesJardin,
Principal Photographer

Peter Gaede,
Illustrator

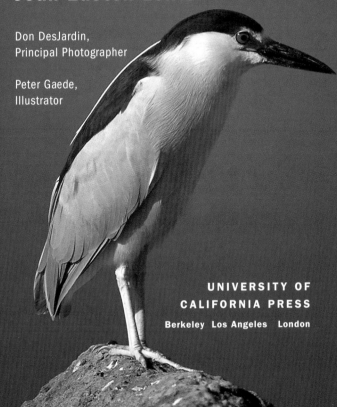

UNIVERSITY OF
CALIFORNIA PRESS

Berkeley Los Angeles London

For Gib

University of California Press, one of the most distinguished university presses in the United States, enriches lives around the world by advancing scholarship in the humanities, social sciences, and natural sciences. Its activities are supported by the UC Press Foundation and by philanthropic contributions from individuals and institutions. For more information, visit www.ucpress.edu.

California Natural History Guide Series No. 84

University of California Press
Berkeley and Los Angeles, California

University of California Press, Ltd.
London, England

© 2006 by The Regents of the University of California

Library of Congress Cataloging-in-Publication Data

Lentz, Joan Easton
    Introduction to birds of the southern California coast / by Joan Easton Lentz.
        p. cm.— (California natural history guides ; 84)
    Includes bibliographical references (p. 293) and index.
    ISBN 0-520-23780-3 (cloth : alk. paper)—ISBN 0-520-24321-8 (pbk. : alk. paper)
    1. Birds—California, Southern. 2. Bird watching—California, Southern.
I. Title. II. Series.
QL684.C2L46 2005
598′.09794′9—dc22

                                                                2004014188

Manufactured in China
10   09   08   07   06
10   9   8   7   6   5   4   3   2   1

Cover photograph: Brandt's Cormorant *(Phalacrocorax penicillatus).* Photograph by Don DesJardin.

The publisher gratefully acknowledges the generous
contributions to this book provided by

the Gordon and Betty Moore Fund
in Environmental Studies
and
the General Endowment Fund of the
University of California Press Foundation.

If a parent wishes to give his children three gifts for the years to come, I should put next to a passion for truth and a sense of humor, love of beauty in any form. Who will deny that birds are a conspicuous manifestation of beauty in nature?

RALPH HOFFMANN, *BIRDS OF THE PACIFIC STATES* (1927)

CONTENTS

PREFACE

Southern California's beaches are world famous. In scenes from movies and television, sun shines on white sands while glamorous people frolic in the blue Pacific. There is not a bird in sight.

This is one view of southern California's coast.

Another entirely different view is drawn from observations of the birds that live there. Behind the stereotype of coastal southern California lies a natural world that still manages to survive. For birds, availability of places to shelter and feed is paramount. Despite grave loss of habitat in the past century, successful restoration projects and heightened public conservation have maintained the region as a top birding destination.

Southern California attracts an unusually rich avifauna, thanks to its habitat diversity, coastal location, and mild climate. From Morro Bay to San Diego, an amazing array of bird species inhabits the nearshore waters, rocky promontories, and tidal mudflats. In some areas, major freeways roar a few yards away and crowded housing tracts mount the hills. Surprisingly, the bird life endures and often even thrives.

On the mudflats of San Diego Bay, a Great Blue Heron stalks minnows. Wandering Tattlers and Black Turnstones scramble over wave-washed rocks at La Jolla. Ducks flaunt their bright colors at coastal lagoons such as Batiquitos and San Elijo. Rails squeeze between the reeds at the marsh at San Joaquin Wildlife Sanctuary. Four species of terns, accompanied by Black Skimmers, scream and hover over a crammed nesting colony at Bolsa Chica. And every spring, thousands of Red-throated and Pacific Loons on their northward migration fly past Point Vicente on the Palos Verdes Peninsula.

Farther north at Malibu Lagoon, flocks of wintering sandpipers gather to rest at the edge of the beach. Brown Pelicans bathe in the protection of the Santa Clara River estuary, while gulls cry from the beach and an Osprey circles overhead. Piedbilled, Horned, Eared, Western, and Clark's Grebes dive in the sheltered waters of Santa Barbara Harbor. And at Montana de Oro in San Luis Obispo County, Black Oystercatchers wield their red bills in defense of seaweed nests on the dark rocks.

Birders and nature lovers seek a different southern California coast, one less familiar, one filled with bird life. Driven by a desire to explore and reconnect with the natural world, these observers search for locations where they can get close to birds, overcoming the challenges of an urban environment. They shrug off noisy freeways and crowded beaches in pursuit of the last wild places where birds live.

This book is an introduction to the birds of the southern California coast. As such, it is designed to familiarize you with the appearance and behavior of most of the common coastal birds, where and when they occur, and how to go about finding them on your own.

ACKNOWLEDGMENTS

I would like to acknowledge the photographers who, in addition to Don DesJardin, contributed their fine work to this book: Ellen Easton, David Koeppel, Peter LaTourrette, Brad Sillasen, and Walter Wehtje.

The following individuals shared their expertise or gave other generous assistance: Karen Bridgers, Charles Collins, Paul Collins, Dave Compton, Don DesJardin, Jon Dunn, Tom Edell, Peter Gaede, Kimball Garrett, Carol Goodell, Robert Hamilton, Marilyn Harding, Marie Holmes, David Kisner, Curtis Marantz, Robert Patton, Cristina Sandoval, Brad Schram, Terri Sheridan, Mike San Miguel, Rich Stallcup, Philip Unitt, and Doug Willick. I am especially grateful to Paul Lehman and Larry Ballard for reading earlier drafts of portions of the manuscript and sharing their suggestions.

I am indebted to the staff at the University of California Press for their encouragement and guidance: Kate Hoffman, Scott Norton, Jenny Wapner, and, especially, Doris Kretschmer (now retired).

Finally, I wish to acknowledge the influence of Ralph Hoffmann, whose book *Birds of the Pacific States* was one of my first field guides, and whose splendid descriptions of birds continue to inspire me.

Joan Easton Lentz
Santa Barbara, California
March 2004

A Bird's-eye View

To get a bird's-eye view of the features of the southern California region, let's follow a common shorebird on a portion of its spring migration route northward from Baja California. Our bird, a Western Sandpiper *(Calidris mauri)* less than 7 inches long, has spent the winter in the big mudflat estuary at Punta Banda, 60 miles south of the Mexican border near Ensenada. It is April, and this male's spring plumage has molted. The black chevrons on his chest show boldly on a white background. His chestnut crown and back feathers gleam.

The sun instills a restlessness in the Western Sandpiper with the lengthening days. Night after night the restlessness builds, and the bird feeds frenetically in preparation for its long journey. But each night is not quite the right one for departure. Meanwhile, hundreds of other Western Sandpipers arrive from points south—Panama, Peru—where they have wintered. After feeding for a day or two on the mudflats, they are set to fly out again.

Our Western Sandpiper seems ready to join these migrant flocks leaving for nesting grounds in Alaska. But the following morning, a thick fog creeps in from the sea, obscuring everything, and the journey must be postponed. Because food is plentiful here, the birds gorge one more day.

At last, the fog dissipates and the evening sunset is clear. A light breeze begins to blow out of the southeast, twilight descends on Punta Banda, and the birds take off. Up with the clouds, flying steadily into the night with a flock of other sandpipers, the Western Sandpiper begins his northward journey.

It might take the Western Sandpiper and his flock a day or two or three to travel the length of the southern California coast on their way north, or it could take only a matter of hours. Speedy Western Sandpipers have been clocked at 45 miles per hour, but most take a more leisurely pace, leapfrogging from one stopover to another.

Dawn finds the Western Sandpipers flying high above San Diego Bay. This is one of several large tidal bays the birds will see en route. The margins of the bays provide excellent feeding spots for shorebirds if the tide is low and the mudflats, rich in invertebrate prey, are accessible. Estuaries, where the larger rivers enter the sea, are key resting and feeding spots, also, but they are scarce in arid southern California. Surprisingly, even rivers that have

Figure 1. Western Sandpiper flying above Morro Bay.

been altered by humans — such as the San Diego River over which the birds now pass, and the Los Angeles River, which lies farther north — prove tempting to the birds if water levels are low.

North of the dark cliffs at La Jolla, the Western Sandpiper crosses a series of coastal lagoons. At San Elijo, Batiquitos, and Buena Vista Lagoons, the mouths of streams form little lakes and marshes. Those that have been conserved or restored become prime targets for the birds, but a stop now would be premature for the flock.

When the sandpipers reach Orange County and pass Laguna Beach, they scan the shoreline for a suitable spot to rest and feed. But where? Thousands and thousands of houses huddle at the coast, march up the hillsides, crowd around the lagoons. Traffic-clogged freeways thread the maze of buildings and surface streets. Our Western Sandpiper spies the inlet of Newport Bay up ahead. The tide is low, and the mudflats of Upper Newport Bay beckon. He wheels sideways and the rest of the flock follows, spiraling downward to land on the muddy shores of the bay. Immediately, the birds begin foraging on the tidal flats, probing the mud for the tiny animals hidden there. Then they may doze, bills tucked beneath their wings, huddled together at the edge of the marsh. At last, rested and well fed, thanks to the providence of Upper Newport Bay, our sandpiper flies onward with the flock.

The blue Pacific Ocean stretches westward to the horizon. In the distance, the birds can see the Channel Islands. The steep ledges of the island bluffs are not good stopovers for sandpipers, who seek the shallow wetlands found on mainland shores. By negotiating a route using landmarks as guideposts, sandpipers rely on generations of instinctual learning to help them navigate the course. They use a multitude of cues, among them the location of bays, estuaries, and headlands, as they make their way up the coast.

Reaching the bulge of land known as the Palos Verdes Peninsula, the Western Sandpiper skirts Point Fermin and Point Vicente, then flies across the wide arc of Santa Monica Bay, passing over Malibu Lagoon on the other side.

Eventually, the sandy beaches of Ventura come into view. Although they look promising, today the beaches are too crowded with people to provide good habitat. Besides, Western Sandpipers are not adapted for the shifting sands and breaking waves of the open beach. The estuary at the Santa Clara River mouth might furnish a welcome refuge, but at the moment it is too deep, forming a lake dammed up by a sandbar on the ocean side.

Onward and northward, driven by the desire to be first on the breeding grounds, the Western Sandpiper follows the southern Santa Barbara coastline and, finally, rounds Point Conception. Here, the character of the coast changes dramatically: waves crash against rocky shores, cliffs tumble straight down to the sea. The northwest wind blows stronger than before.

Exhausted and hungry, each bird in the flock must draw on inner reserves. With their migratory urge primed, the birds cannot rest yet. They tighten their formation, wings beating faster and faster, flying low, hugging the shore to avoid the force of the wind. They circle over the Santa Maria River mouth and linger for a moment. But the water in the estuary has backed up into a lagoon. No mudflats here; the birds must press on.

At last, the inlet of Morro Bay appears in the distance. The dome of Morro Rock shines in the afternoon sun. The expansive mudflats glisten, the tide is falling, and already hundreds of other

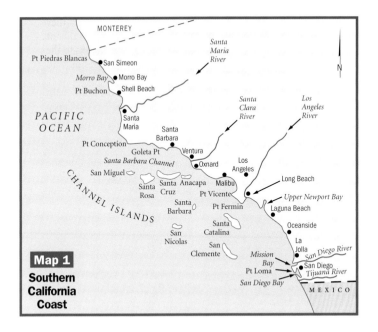

Map 1
Southern California Coast

shorebirds have congregated to feed there. The sandpipers, their strength depleted, descend swiftly and land at the eastern edge of Morro Bay. They feed on the margins of puddles left by the receding water. Here, they will stock up, rest, and restore themselves for the upcoming leg of the trip.

Tomorrow or the next day, the Western Sandpiper will continue his passage north, aiming for the tidal mudflats of Monterey or San Francisco Bay. By mid-May, he will have arrived with millions of other Western Sandpipers to nest at the Yukon-Kuskokwim Delta in western Alaska.

Boundaries of the Coastal Strip

Now let us take a closer look at the territory over which our hypothetical sandpiper has flown: the narrow strip along the southern California coast where the land meets the sea.

The waters up to a mile offshore constitute the western limit of the region, and the eastern limit may be up to one or two miles inland, as when the edges of some bays, inlets, or marshes are not located at the immediate coastline. Though somewhat arbitrary, the concept of a coastal strip as an important corridor for birds is a useful one, directing attention to the narrow belt where many of southern California's birds, and much of its human population, are concentrated.

The northern and southern limits of the southern California coastal strip are easily defined. To the south, the region ends at the U.S.-Mexico border. To the north, the San Luis Obispo–Monterey County line works well as a boundary. The bird life north of the San Luis Obispo–Monterey County line shares an affinity with that of northern California, while that south of the line has characteristics consistent with more southerly species. Of course, northerly species sometimes venture south, especially along the portion of the coast north of Point Conception. Similarly, southerly species often cross the U.S.-Mexico border. An exciting mix of birds from the north and the south makes "anything's possible" a good motto for southern California birding.

The Channel Islands are not technically part of the coastal strip, but they are mentioned in the text where appropriate, as, for example, when several species observed along the mainland coast use the islands for nesting.

Climate

Southern California's climate is important to the abundance of its bird life. Our mild winters attract numerous birds from northern and inland regions, fleeing cold weather and frozen conditions to shelter on southern California's coast. Winter and summer temperatures are moderate along the coastal strip because ocean waters—varying only a few degrees from winter to summer—influence temperature on the nearby land, keeping it cooler in summer and warmer in winter.

Southern California has a Mediterranean climate, which we share with only 1 percent of the earth's land area. Located on the western border of a continent between 30 and 45 degrees latitude, these areas have cool, wet winters and warm, dry summers.

On balance, our climate is semi-arid, due to the lack of precipitation. Winter storms occur from November through March, usually followed by a long dry spell. Average annual rainfall decreases from north to south: in San Luis Obispo it is approximately 23 inches, in Santa Barbara 18 inches, in Los Angeles 15 inches, and in San Diego 10 inches.

Habitats

One of the ways of understanding the birds of a region is to break the region into habitats. Habitats are living and feeding areas where one species of bird is more likely to be found than another. For example, the Western Sandpiper is suited for feeding at tidal mudflats, but it is unable to forage when the mudflats are flooded and the water is too deep.

Where the land meets the sea, a mosaic of habitats attracts a variety of bird species. A cross section of the southern California coastal strip contains four major regions: ocean, seashore, coastal wetland, and coastal terrace. For a more accurate description, the regions can be further divided into eight habitats for birds: nearshore waters, rocky shore, sandy beach, tidal mudflat and lagoon, salt marsh, freshwater marsh, grassland, and coastal sage scrub.

Ocean

From Point Conception to the Mexican border, the big open bay of the Pacific Ocean is known as the Southern California Bight. Although spelled differently from the word "bite," it means about the same: a bend or curve in the coastline. Looking at the way the California coastline curves south of Point Conception, you can imagine a giant fish grabbing a big bite out of the land.

Most of the ocean covered by this book lies within the Southern California Bight, but the waters north of Point Conception lie outside it. Conditions of waters along the southern California coast are complex, being a mixture of the currents and conditions north of Point Conception and those to the south. The California Current, which flows southward along the coast, brings its cool waters past Point Conception, then flows outside the Northern

Figure 2. Nearshore waters: (left to right) Double-crested Cormorant, Western Gull, Common Loon, Brown Pelican, Surf Scoter, Western Grebe, and Red-throated Loon.

Channel Islands. When the current curls back toward the coast, it becomes part of a giant, counterclockwise eddy that mixes the waters of the north with those of the south flowing up from Mexico.

Nearshore Waters

Birds of nearshore waters forage in waters that are close enough to shore that birders can watch them with a spotting scope or a pair of binoculars. Nearshore waters include the open ocean, as well as the relatively shallow waters of kelp beds, larger bays, smaller embayments, and human-made harbors. The largest species of algae, the giant kelp *(Macrocystus pyrifera)* grows here.

Red-throated, Pacific, and Common Loons *(Gavia stellata, G. pacifica,* and *G. immer)*, which have wintered in Baja California, fly northward along the coast in spring migration, stopping to feed on fish along the way. Western and Clark's Grebes *(Aechmophorus occidentalis and A. clarkii)* float in squadrons, their long, black-and-white necks graceful above the winter swell. Eared Grebes *(Podiceps nigricollis)*, plainer than their swanlike cousins, swim in harbors closer to shore. Brown Pelicans *(Pelecanus occidentalis)* plunge-dive from above for fish, shadowed by Heermann's Gulls *(Larus heermanni)* hoping to steal their catch. Brandt's, Double-crested, and Pelagic Cormorants *(Phalacrocorax penicillatus, P. auritus,* and *P. pelagicus)* ply the waters, diving

from the surface to chase prey underwater. Surf Scoters *(Melanitta perspicillata)* tug at mollusks clinging to rocks and pilings, or lounge in shallow bays. The white forms of gulls and terns circle and dive, their wild cries rising above the wind and waves.

When ocean waters are ripe with zooplankton (tiny organisms that float in the ocean), Sooty Shearwaters *(Puffinus griseus),* mass in flocks close to shore. They alight on the water, sip the nutritious plankton, then take off again. Great Blue Herons *(Ardea herodias)* and Great Egrets *(A. alba)* sometimes visit the kelp beds, standing as solitary silhouettes on the kelp itself or on pieces of driftwood nearby as they hunt for fish.

The best places to see birds in this habitat are from the ends of piers and breakwaters, from coastal promontories, and from boats. Examples include La Jolla, Newport Pier, Point Vicente Fishing Access, Point Dume, Mugu Rock, Goleta Point, Goleta Beach Pier, the Santa Maria River estuary (Guadalupe Dunes County Park), Montana de Oro State Park, and Cayucos Pier.

Seashore

The seashore, or littoral zone, is the land at the ocean's edge. Where the coast is formed of hard rocks, the land resists erosion and forms rocky headlands. At the base of these headlands, a rocky intertidal habitat harbors a rich community of invertebrates such as worms, crabs, and mussels. Between the headlands are beaches, where sand is deposited by waves.

Rocky Shore

Birds inhabiting the steep cliffs and headlands of the rocky shore habitat are specialists in prying, turning, and snatching at the intertidal creatures that cling to rocky surfaces. Black Oystercatchers *(Haematopus bachmani)* chisel mussels open with their long red bills. Black Turnstones *(Arenaria melanocephala)* and Ruddy Turnstones *(A. interpres)* creep over the rocks, picking at the detritus left by the last wave, and overturning shells to find amphipod crustaceans. Before the next wave crashes, they fly up in a shower of black and white wings, then settle again. Wandering Tattlers *(Heteroscelus incanus)* and Surfbirds *(Aphriza virgata)* revel in the surf spray—they like the wettest rocks, which they search for small crustaceans. Brandt's, Double-crested, and Pelagic Cormorants and Pigeon Guillemots *(Cepphus columba)* nest atop cliffs or on shelves of the rocky shoreline. Also choosing remote sea cliff ledges for nesting are the Common Raven *(Corvus corax)* and the Peregrine Falcon *(Falco peregrinus)*.

When extreme minus tides occur, other species of birds join the rocky shorebirds at exposed tide pools. Sandpipers, herons, egrets, and gulls explore the rocks before the tide turns. Even landbirds such as the American Crow *(Corvus brachyrhynchos)*, Euro-

Figure 3. Rocky shore: (left to right) Black Turnstone, Ruddy Turnstone, Brandt's Cormorant, Surfbird, Wandering Tattler, Black Oystercatcher, Peregrine Falcon, and Pigeon Guillemot.

pean Starling *(Sturnus vulgaris),* and Brewer's Blackbird *(Euphagus cyanocephalus)* venture onto the rocks at low tide.

Rocky shores are more prevalent along the coast north of Point Conception and on the Channel Islands. South of Point Conception, rocky shores are disjunct, separated by long stretches of sand or low bluffs. Throughout, human-made breakwaters and jetties provide some of the same conditions found on rocky shores.

Examples of this habitat are found at Sunset Cliffs, La Jolla, Heisler Park, Royal Palms State Beach, Playa del Rey jetties, Ventura Harbor jetties, the Santa Barbara Harbor breakwater, and the cliffs at Shell Beach.

Sandy Beach and Dunes

Where there are no sea cliffs along the coast, the winds and currents carry sand, creating beaches. On a map, you can see that many headlands along the coast curve to the south, protecting a bay or river to the east. Where the river meets the ocean (at an estuary) it is often blocked, or nearly so, by a sandbar across its mouth, which separates the river from the ocean. Coastal sandbars and beaches are formed when particles torn from the head-

Figure 4. Sandy beach and dunes: (left to right) Snowy Plover, Marbled Godwit, Whimbrel, Sanderling, Black-bellied Plover, and Willet.

lands are transported by currents and dropped as they meet the quieter waters of the bay or estuary.

As the waves break along the beach, plump little Sanderlings *(Calidris alba)*, accompanied by Marbled Godwits *(Limosa fedoa)* and Willets *(Catoptrophorus semipalmatus)*, hurry up and back, seeking prey in the ever-shifting substrate of the wet sand. Other shorebirds, such as Black-bellied Plovers *(Pluvialis squatarola)* and Whimbrels *(Numenius phaeopus)*, search the wrack above the high tide line. A large flock of gulls—Mew *(Larus canus)*, Ring-billed *(L. delawarensis)*, California *(L. californicus)*, Western *(L. occidentalis)*, and a few Glaucous-winged *(L. glaucescens)* —rest and preen on the sand up by the parking lot. Black Phoebes *(Sayornis nigricans)* and American Pipits *(Anthus rubescens)*, although they are not shorebirds, are attracted to sandy beaches by the insect life in the piles of kelp.

Most birds that visit the sandy beach in winter are largely gone by May, having migrated to breeding grounds farther north. They are replaced by the only two bird species that nest in the sandy beach habitat: the California Least Tern *(Sterna antillarum browni)* and the Western Snowy Plover *(Charadrius alexandrinus nivosus)*, both now protected as endangered or threatened species. With monitoring efforts, the tern and the plover are slowly recovering on certain of southern California's beaches.

Examples of this habitat are at Tijuana Slough National Wildlife Refuge, Doheny State Beach, Leo Carrillo State Beach, Ormond Beach, Sands Beach (Coal Oil Point Reserve), the Santa Maria River estuary (Guadalupe Dunes County Park), and the dunes west of Oso Flaco Lake.

Coastal Wetland

An estuary, where a river meets the ocean, creates several important habitats for birds. Estuaries are nurseries for fish. The adults spawn in the shelter of the calm waters where the young have a better chance to survive. Numerous marine invertebrates live in burrows or tubes in the mud.

Birding along the coast in tidal areas necessitates an awareness of the tides. A dry mudflat can become a shallow bay in a few hours. Along the southern California coast are two low and two high tides each day. They cycle through a 25-hour period, so each tide is about an hour later than the one the day before. Twice a month, extreme tides (spring tides) are caused by the position of the sun, earth, and moon. Less extreme tides (neap tides) are those with a small rise and fall. In May and June and in November and December (around the solstices) the tides are very high and very low.

Figure 5. Tidal lagoon: (left to right) Black-crowned Night-Heron, Osprey, Red-breasted Merganser, Bufflehead, Ring-necked Duck, Green-winged Teal, and Spotted Sandpiper.

Estuarine Mudflats, Tidal Lagoons, and Brackish Waters

Where a river enters the sea at the mouth of an estuary, there is a constant ebb and flow of freshwater and seawater, a shifting boundary that moves with the tides. The twice daily flushing of the tides creates two habitats: tidal mudflat and tidal lagoon.

Tidal mudflats are exposed as the tide goes out. The marine animals that live in tubes or burrows in the mud are plankton feeders, dependent on the tides for daily nourishment. When the tide falls, these animals—worms, shrimps, clams, snails, and crabs—retreat into their burrows. But they cannot escape the probing bills of the numerous birds that gather at this accessible feast. From the tallest, such as the Long-billed Curlew (*Numenius americanus*) and the Marbled Godwit, to the smallest—the Least Sandpiper (*Calidris minutilla*)—sandpipers search the mudflats for organisms. In addition, American Avocets (*Recurvirostra americana*), Black-necked Stilts (*Himantopus mexicanus*), Black-bellied and Semipalmated Plovers (*Pluvialis squatarola* and *Charadrius semipalmatus*), Killdeer (*Charadrius vociferus*), and herons and egrets join the throngs of birds at the estuary "café."

When the tide is in, a shallow bay or lagoon forms; the water here can be several feet deep. (The lagoon may persist if the estuary channel to the sea is blocked by a sandbar.) Grebes, pelicans,

cormorants, geese, ducks, and coots take the place of the shore-birds. In the deeper water, Lesser Scaup *(Aythya affinis)* and Ring-necked Ducks *(A. collaris)* float and dive. Redheads *(A. americana)* and Canvasbacks *(A. valisineria)* may be scattered among them. The small black-and-white Buffleheads *(Bucephala albeola)* dive, then pop up for air. A Red-breasted Merganser *(Mergus serrator)* emerges with a minnow squirming in its bill. Black-crowned Night-Herons *(Nycticorax nycticorax)* sit hunched over, dozing away the daylight hours. On the margins of the lagoon, where shallower waters abut the marsh vegetation, Cinnamon and Green-winged Teal *(Anas cyanoptera* and *A. crecca)* nuzzle for seeds and plant material in the mud.

Raptors (birds of prey) are drawn to the tidal mudflats and lagoons by the host of birds feeding there. The Peregrine Falcon *(Falco peregrinus),* one of the most successful tidal mudflat predators, unerringly picks off a shorebird and carries it away in sharp talons. Ospreys *(Pandion haliaetus),* on the lookout for fish, circle high above tidal lagoons or perch on pilings in the middle of the water.

With the constant inundation of saltwater, most plants are unable to survive at the lowest levels of an estuary. The bright green eel-grass *(Zostera marina),* however, grows from the low-tide level to depths of 20 feet. It shelters diatoms and small invertebrates—and is a favorite food of wintering and migrating Brant *(Branta bernicla).*

Cord grass *(Spartina foliosa)* is another important plant of the lower estuary. It excretes excess salt by means of special glands. Cord grass also has hollow passageways in its leaves and roots so air can circulate throughout the plant even when it is underwater. This plant contributes to the health of the estuary ecosystem by trapping sediments and detritus, which are then devoured by worms, snails, and crustaceans. They in turn are consumed by fishes and birds. Waste material, replete with nitrogen, is carried out to sea on the next low tide.

The health of an estuary depends upon the amount of freshwater coming in and the degree to which the tidal action circulates saltwater. One of the problems with southern California estuaries is that reduction of freshwater input, siltation, and human encroachment keep the barrier sandbar from being breached for months at a time.

The best places to view tidal mudflats and lagoons are at major estuaries and bays. Examples are found at South San Diego Bay, Upper Newport Bay, Bolsa Chica, Malibu Lagoon, the Santa Clara River estuary, Goleta Beach County Park, the Santa Maria River estuary, and Morro Bay.

Figure 6. Salt marsh: (left to right) Belding's Savannah Sparrow, Forster's Tern, Long-billed Curlew, American Avocet, Snowy Egret, Belted Kingfisher, and Clapper Rail.

Salt Marsh

The salt marsh borders the tidal mudflats and lagoon on the landward side of an estuary. Pickleweed (*Salicornia* spp.) is the low, gray-brown groundcover that dominates. Like cord grass, pickleweed is very salt tolerant. The roots of pickleweed support the dikes and banks of the salt marsh, keeping drainage channels open.

Approximately 90 percent of coastal southern California's salt marsh habitat has disappeared, making it one of the rarest habitats in the world. Although only a few species of birds use the salt marsh as a nesting spot, two of them, the Light-footed Clapper Rail *(Rallus longirostris levipes)* and the Belding's Savannah Sparrow *(Passerculus sandwichensis beldingi),* are listed as endangered. Occasionally, the "clappering" sound uttered by the Clapper Rail can be heard over the expanse of the salt marsh; however, seeing this secretive species is another matter. In contrast, the Belding's Savannah Sparrow perches up quite readily on the pickleweed to sing its spring song.

Migrating and wintering shorebirds roost in the salt marsh when the mudflats are flooded. Belted Kingfishers pick a stake or low bush in the salt marsh, then fly out to fish by hovering above the lagoon. In the hidden pools, American Avocets sieve the water through their up-turned bills as they sidle along.

Examples of salt marsh habitat are seen at Tijuana Slough

Figure 7. Freshwater marsh: (left to right) Red-winged Blackbird, Barn Swallow, Pied-billed Grebe, American Coot, Northern Shoveler, Ruddy Duck, Great Blue Heron, and Common Yellowthroat.

National Wildlife Refuge, San Elijo and Batiquitos Lagoons, Upper Newport Bay, Bolsa Chica, Mugu Lagoon, Carpinteria Salt Marsh Nature Park, Devereux Slough (Coal Oil Point Reserve), and Morro Bay (Morro Bay State Park marina area).

Freshwater Marsh

Freshwater marshes occur at the upstream end of estuaries, and along the margins of lakes, ponds, and sloughs. On the southern California coast, these areas are few and far between. Where freshwater marshes remain, they are packed with wildlife. They attract grebes, herons and egrets, geese, ducks, rails, coots, and several species of perching birds.

Most freshwater marshes contain reedlike plants that grow in water-saturated soil. Rushes (*Juncus*), bulrushes or tules *(Scirpus)*, sedges *(Cyperus)*, and cattails *(Typha)* grow in thick clumps in freshwater marshes. The cattails are easily spotted; they have flower spikes that look like long brown hot dogs on sticks. Tules are usually not as tall as cattails, and their flowers are clusters of reddish brown blossoms at the tops of the stems. These plants are

used for nesting materials, perches, and, especially, for cover by birds inhabiting the freshwater marsh.

Flotillas of Northern Shovelers *(Anas clypeata)*, Mallards *(A. platyrhynchos)*, and Ruddy Ducks *(Oxyura jamaicensis)* float on freshwater ponds. Pied-billed Grebes *(Podilymbus podiceps)* negotiate the edges where they hide among the tules. American Coots *(Fulica americana)*, clumsy-looking blackish birds with white shields on their bills, dive for plant matter with a splash. Marsh Wrens *(Cistothorus palustris)* and Common Yellowthroats *(Geothlypis trichas)* scoot up and down the tule stems, their ebullient songs resonating over the water. Other denizens of the freshwater marsh, such as Virginia Rails *(Rallus limicola)* and Soras *(Porzana carolina)*, are vocal, too, but seldom show themselves. In spring, the male Red-winged Blackbird *(Agelaius phoeniceus)* is anything but quiet as he puffs out his red shoulder patches and sings from the tops of the tules.

Cliff and Barn Swallows *(Petrochelidon pyrrhonota* and *Hirundo rustica)* swoop in the air over the marsh, feeding on gnats and midges. A Green Heron *(Butorides virescens)* waits, camouflaged among the reeds. If the marsh has enough open water, a group of American White Pelicans *(Pelecanus erythrorhynchos)* may visit.

Figure 8. Grassland: (left to right) American Kestrel, White-tailed Kite, Western Meadowlark, Northern Harrier, Say's Phoebe, and Red-tailed Hawk.

Examples of freshwater marsh habitat are found at Buena Vista Lagoon, San Joaquin Wildlife Sanctuary, Ballona Freshwater Marsh, Ventura Water Treatment Plant/Wildlife Ponds, Andree Clark Bird Refuge, Lake Los Carneros, and Oso Flaco Lake. (Note: Freshwater marshes are often more accessible in locations farther inland from the immediate coast, such as at San Joaquin Wildlife Sanctuary and Lake Los Carneros.)

Coastal Terrace

If you walk inland from coastal wetlands, or climb up the cliffs above rocky shores, one of two plant communities greets you: grassland or coastal sage scrub. Often there is a mixture of the two. Bird species in these two habitats cannot be considered strictly coastal, because they are also found in bird communities farther inland. Therefore, only the most common species (out of a great many) are emphasized below and in the species accounts.

Grassland

Grassy areas are rare along the southern coast. In Santa Barbara and San Luis Obispo Counties, a few coastal terraces covered with grassland remain. Most of the open areas that have not been

urbanized have been given over to agriculture or pastureland. Trees are scarce, except for the occasional clump of willow or eucalyptus.

A White-tailed Kite *(Elanus leucurus)* hovers high above the pasture, its keen eyes watching for prey hidden in the grass. A Red-tailed Hawk *(Buteo jamaicensis)* watches the grasses from its perch on the eucalyptus tree. The American Kestrel *(Falco sparverius)* and Northern Harrier *(Circus cyaneus)* hunt for insects and small rodents in open country, perching on utility poles or fenceposts. At night, Barn Owls *(Tyto alba)* glide silently over the fields.

Western Meadowlarks *(Sturnella neglecta)*, one of the species that actually nests in the grass, gather in large feeding flocks during fall and winter. Their bright yellow chests shine against the green background of new grasses. Many other birds use the grasslands for foraging during fall and winter months but leave in the spring to nest elsewhere. Say's Phoebes *(Sayornis saya)* perform flycatching acrobatics from the tallest weed stalks. The streaky Savannah Sparrows *(Passerculus sandwichensis)* perch on fencewires. Big groups of Red-winged Blackbirds and European Starlings *(Sturnus vulgaris)* descend on the grasslands, especially if cattle roam the fields.

Examples of coastal grassland are at More Mesa (in Goleta), Estero Bluffs Trail north of Cayucos, and fields along California Route 1 on the San Simeon coast.

Coastal Sage Scrub

Growing right to the upper margins of the salt marsh or on the edges of coastal grasslands, an association of plants called coastal sage scrub thrives. Many of its plants are aromatic and have soft, flexible leaves and stems. In fall and winter, birds depend on the seeds, berries, and flowers borne by this plant community. In the breeding season, the dense branches of the shrubs provide nesting spots for birds and help conceal them from predators.

Coastal sage scrub is named for the variety of sages that dominate this plant community. California sagebrush *(Artemisia californica)* is a soft, pungent-smelling plant with feathery leaves. If you brush past its foliage, you will take away the smell that, more than any other, epitomizes the coastal plants of southern California. Other sages are black sage *(Salvia mellifera)* and purple sage *(Salvia leucophylla)*.

Song Sparrows *(Melospiza melodia)* hop through the tangled vines of the moister sections of the scrub. In the shaded arroyos, they are joined by Common Yellowthroats. The drab California Towhee *(Pipilo crissalis)* and its bright cousin, the Spotted Tow-

Figure 9. Coastal sage scrub: (left to right) Song Sparrow, American Crow, California Towhee, White-crowned Sparrow, Red-shouldered Hawk, and Spotted Towhee.

hee *(Pipilo maculatus),* are familiar sights. The California Towhee feeds on seeds beside the trail. The Spotted Towhee sneaks in and out near the base of the shrubs, seldom giving you a good, clear view. In fall and winter, White-crowned and Golden-crowned Sparrows *(Zonotrichia leucophrys* and *Z. atricapilla)* sing their wistful tunes. North of Point Conception, the resident Nuttall's subspecies of the White-crowned Sparrow *(Z. l. nuttalli)* breeds in coastal sage scrub.

Many resident birds of coastal sage scrub are rather dull in color. Most have relatively short wings; they are weak fliers, for they do not need to travel far when moving from one bush to another. Many have long tails, which help them navigate up and down through dense shrubbery.

A detailed description of all of the common bird species of coastal sage scrub is beyond the scope of this book. Furthermore, most of the species are primarily inhabitants of the chaparral plant community, which exists farther inland. One bird worth mentioning, however, is a scarce resident in coastal sage scrub from Ventura County southward: the California Gnatcatcher *(Polioptila californica).* A tiny gray bird, the California Gnatcatcher has been recently designated as a threatened species by

the federal government, as a result of the imminent destruction of portions of its nesting habitat along the coast.

The best time to study birds in coastal sage scrub is February through April. Wintering species are still present, and residents are starting to nest. Most scrub plants are in bloom then, their beautiful blossoms and fragrances a bonus for birders.

Examples are seen at visitor center trails at Tijuana Slough National Wildlife Refuge, trails at San Elijo and Batiquitos Lagoons, Ocean Trails (Palos Verdes Peninsula), More Mesa in Goleta, and Cerro Cabrillo trailhead (Morro Bay State Park).

Birding Basics: What to Look For

Body Features

It is helpful for a beginner to become familiar with the shape and structure of each of the common bird families. For this reason, a silhouette representing the bird family is placed at the start of the discussion on each family in the species accounts.

Some generalizations about bird shapes are easy to make. Birds with long legs are waders and spend time in watery environments, where long legs are an advantage. Predators have sharp beaks and hooked claws for grabbing and killing prey. Birds with a large wingspan are capable of extended periods of time in the air.

Notice the bird's size first. Perhaps the best strategy is to compare the size of an unfamiliar bird with the size of one you already know. For example, "small" birds might be the size of a Song Sparrow *(Melospiza melodia),* a common bird of coastal wetlands. Or, if you are familiar with a certain group of birds, try to place the unfamiliar candidate into a range, such as a medium-sized gull or a medium-sized shorebird.

Next, observe the shape of the bird's bill. Bill shapes give major clues to bird families, because variations in bill size and shape signify certain feeding behaviors. Because bird families often comprise species with similar feeding behaviors, bill structure is important. Sparrows have small, conical bills efficient at opening and cracking seeds. Flycatchers have slender bills that enable them to snare insects on the wing. Members of the sandpiper family generally have long, slender bills for probing in mud or sand. Furthermore, within each family slight differences in bill shape are often enough for correct identification of species. Among loons, cormorants, ducks, gulls, and terns, careful observation of bill shape can be diagnostic.

Wing and tail shapes are important to the overall impression of a bird. Ravens and crows are both large, dark birds, but in flight the raven's tail is wedge shaped coming to a point, whereas the crow's is more rounded. Some of southern California's swallows are difficult to distinguish in flight, but the Barn Swallow (*Hirundo rustica*) is easily spotted by its forked tail.

Wing shapes and wing beat patterns are helpful when the bird is silhouetted against the sky in flight. Terns have pointed wings sharply bent at the "wrist," gulls do not. To tell a shearwater in the distance from a gull, look for the rapid wing beats alternating with stiff-winged glides of the shearwater, as opposed to the steady flapping of a gull.

Plumage

Look at the color and pattern of the bird's feathers. These "field marks" include the color or contrast of the bird's cap, breast, bellyband, and outer tail feathers; the presence of eye rings, wing bars, eye stripes, rump patches, or tail spots; and any spots, scalloping, or streaking on the head or breast.

When describing a bird, begin with its head. Important field marks are often concentrated in the head pattern, and diagnostic head markings help separate one species of bird from another. For example, male Blue-winged, Cinnamon, and Green-winged Teal (*Anas discors, A. cyanoptera,* and *A. crecca*) all have colorful, distinctive head markings. Eared and Horned Grebes (*Podiceps nigricollis* and *P. auritus*) are separated in winter plumage by their cheek patches.

After observing the head, look at the rest of the bird for prominent field marks. The color of the wing tips (primaries) on gulls can be critical. The red patch on the wing of the Red-winged Blackbird (*Agelaius phoeniceus*) is found on its lesser coverts. When flying, the Northern Harrier (*Circus cyaneus*) always shows a prominent white patch on its rump.

All plumage changes in birds are achieved by molt. Molt refers to the periodic replacement of feathers. Most adult birds molt at

Figure 10. Topography of a bird: Heermann's Gull.

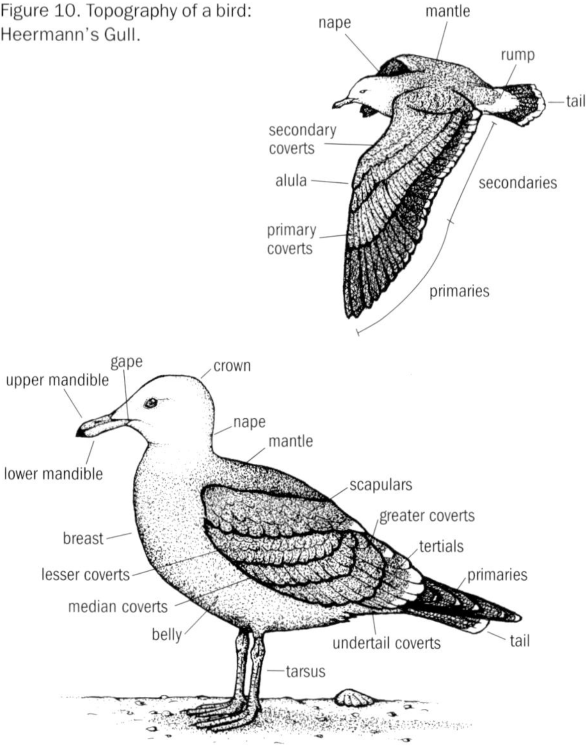

least once a year, some even two or three times: feathers are so essential that birds need to replace them when they are worn. In addition, birds need different kinds of feathers at different times of their lives. The males and females of most seabirds, such as loons, grebes, and cormorants, and most shorebirds, such as sandpipers and plovers, have a more vivid breeding plumage (in spring) and a less colorful nonbreeding plumage (in fall and winter). Among many landbirds or perching birds, such as the Common Yellowthroat *(Geothlypis trichas)* or the Red-winged Blackbird, males have a more colorful plumage than females.

In the nest, a chick wears a soft, fluffy cover of down. When a young bird has its first coat of true feathers, it is termed a juvenile.

The bird may molt one or more times before it completes imma-ture or subadult plumages and assumes adult plumage. Once the bird reaches adulthood, it usually has a winter or nonbreeding plumage (basic plumage), followed by a more colorful spring or breeding plumage (alternate plumage). From then on, an adult bird undergoes basic and alternate plumages throughout its life-time, typically tied to the annual cycle of the seasons.

The identification of gulls and shorebirds often depends upon a birder's ability to correctly determine the age of the bird in question. Juvenile birds have a bright, fresh look to them. In late summer and early fall, their pale feather edgings have not worn off yet, giving the young birds a pristine plumage pattern that experienced birders learn to recognize. As an example, the easiest time to tell a Long-billed Dowitcher *(Limnodromus scolo-paceus)* from a Short-billed Dowitcher *(L. griseus)* is when both are in juvenal plumage during fall migration.

Behavior

Some birds can be identified by their behavior alone. Ask yourself this question: What was the bird doing when I saw it? Birds spend most of their time hunting for food. Foraging behavior, which is frequently unique to a particular family or even a particular species, is a clue to a bird's identity.

Does the bird hover in a stationary position in midair over a field? Although several birds of prey employ this hunting tech-nique from time to time, it is most characteristic of the White-tailed Kite *(Elanus leucurus)* and the American Kestrel *(Falco sparverius)*. Others, like the Northern Harrier, fly low over open country, rocking from side to side. The Turkey Vulture *(Cathartes aura)* has this same tilting flight pattern, but it usually remains high in the sky. Among shorebirds, feeding behavior is related to the length of the bird's bill. In addition, some shorebirds have special tail movements and postures. The Spotted Sandpiper *(Actitis macularia)* and the Wandering Tattler *(Heteroscelus in-canus)* bob their hind ends constantly when searching for food on mudflats and rocks. Spotted and California Towhees *(Pipilo maculatus* and *P. crissalis)* scratch vigorously in the dry leaves of coastal sage scrub—forward and backward feet together—to uncover seeds.

Flight behavior not connected with foraging is worth noticing, too. Loons and cormorants both have rapid wing beats in flight. Loons, however, have a peculiar "bowed" look, with head and tail held lower than the middle of their bodies, while cormorants hold their bodies level or tilted slightly upward and forward in flight.

Many other types of behavior distinguish bird species. Field guides characterize birds as secretive (rails), aggressive (crows), or gregarious (shorebirds). All are valid descriptions of a multitude of behaviors that help an observer arrive at the identity of a bird species.

Vocalizations

Often a bird's vocalizations are the first and only clue you need to correctly determine its species. A crucial part of birding is heightened awareness of details: open your ears to bird sounds. By expanding your powers of observation, and carefully listening to the sounds birds make, you can learn to recognize bird calls. Although this skill is not as critical at the coast as it is in forested habitats, where birds are often hidden, it is an aid to bird identification wherever you are. Many birds that are too distant to be positively distinguished by sight, such as those on a mudflat, can be quickly identified if they vocalize. Shorebirds, terns, marshbirds, and a variety of passerines (perching birds) along the coast can be distinguished by the sounds they make. The loud, rattling call of the Belted Kingfisher *(Ceryle alcyon)*, the upslurred cry of the Black-bellied Plover *(Pluvialis squatarola)*, the alarm call of the Willet *(Catoptrophorus semipalmatus)*, the "cree-creet?" of the Western Grebe *(Aechmophorus occidentalis)*, the "clappering" sound of the Light-footed Clapper Rail *(Rallus longirostris levipes)*, the thin wheeze of the Belding's Savannah Sparrow *(Passerculus sandwichensis beldingi)*, and the croak of the Common Raven *(Corvus corax)*—these are the sounds of the coast. On the other hand, some birds, especially those that winter here, are silent; loons, herons, pelicans, cormorants, and some ducks fall in this category.

Perching birds that live on coastal sage scrub slopes are very vocal. They sing from the tops of bushes or communicate with each other by "chip" calls in the dense undergrowth. At the edge of a freshwater marsh, where birds are well hidden by cattails and tules, the ability to assign bird calls by species is especially helpful—the Pied-billed Grebe *(Podilymbus podiceps)*, Sora *(Porzana carolina)*, and Virginia Rail *(Rallus limicola)* make loud calls. Listening to a marsh is as good as creeping through it—better to let your ears do the "seeing" and leave the birds undisturbed.

Habitat

Habitat is the set of specific environmental conditions that must be met for the survival of a bird. Weather, vegetation, altitude, freshwater versus saltwater, human presence or lack of it—all these characterize a bird's habitat.

Most birds have a fairly well defined geographical range in which they normally occur at certain times of the year. Usually a species occupies a breeding range to nest and raise young, then moves to a wintering or nonbreeding range for the rest of the year. A bird's distribution pattern shows where it can be found at certain seasons. Knowing the distribution of a species in any given region helps define a bird's seasonal abundance. Along the southern California coast, some birds are permanent residents, some are summer residents, some are fall and winter visitors, and some pass through in spring and fall migration. Many fall into more than one of these categories.

Within its general range, each species lives by preference in a certain habitat. A bird's habitat is usually dictated by its food requirements. Some birds are more adaptable than others when it comes to habitat. For example, the Black Oystercatcher *(Haematopus bachmani)* is restricted to rocky shores, where it pries open shellfish with a specialized bill. It is rarely found away from this habitat. On the other hand, the Willet ventures from the sandy beach to the rocky shore to the tidal mudflats in search of prey.

In short, knowledge of a bird's habitat preferences is critical to locating it. To use extreme examples, birds that forage in ocean waters are not likely to be found in coastal sage scrub, and birds suited to a freshwater marsh are not likely to be found on rocky shores.

Taxonomy

Scientists arrange all organisms in a system of classification called taxonomic order. The organisms are grouped together depending upon their degree of similarity. For example, the taxonomy of a California Least Tern *(Sterna antillarum browni)* appears in table 1.

TABLE 1. Taxonomy of the California Least Tern

KINGDOM	Animalia (animals)
PHYLUM	Vertebrata (animals with spinal cords)
CLASS	Aves (birds)
ORDER	Charadriiformes (shorebird, gull, and alcid families)
FAMILY	Laridae (gulls, terns, and skimmers)
GENUS	*Sterna* (closely related terns)
SPECIES	*antillarum* (Least Tern)
SUBSPECIES	*browni* (California Least Tern)

This system, used worldwide by scientists today, originated in 1758 when Carolus Linnaeus, a Swedish naturalist, published *Systema Naturae*. In the Linnaean system, each organism has a Latin name in two parts: the first is the name of the genus, and the second is the species epithet. Generic names, *Sterna* in this case, are capitalized, whereas species names, *antillarum* here, are not. In some cases, a third Latin name indicates a subspecies. Subspecies are populations, or races, of birds within a given species that are subtly different in ways that may or may not be obvious in the field. Now that molecular studies have enabled a closer understanding of genetic makeup, many subspecies may one day be considered full species and vice versa.

The arrangement of order, family, genus, species is based on the presumed evolutionary development of each group of birds. The most primitive order—in North America, the geese, ducks, and swans (Anseriformes)—is listed first, and the most advanced—the perching birds, or passerines (Passeriformes)—comes last. Bird checklists, as well as many standard field guides, are organized according to the currently accepted taxonomic system.

Ecological Types

Groups of birds are often referred to in general terms, such as landbirds or waterbirds. The terms have their origins in the ecological habitats that these birds occupy. The terms are not a formal part of the Linnaean classification system but prove convenient when discussing groups of bird families.

Waterbirds are those species found near water, and they can be further divided into seabirds, marshbirds, waterfowl, and shorebirds. Seabirds are species living much of their life near or over the sea, such as loons, shearwaters, pelicans, cormorants, gulls, and terns. Marshbirds are those waterbirds that live in salt or fresh-water marshes: herons, egrets, rails, and ibises. The term water-fowl refers to all the species of geese and ducks, and the term shorebirds refers to species that live near water along the coast, such as plovers, oystercatchers, stilts, avocets, and sandpipers.

Landbird species are in the minority along the coast. Land-birds are those species found largely on or over land, such as hawks, falcons, flycatchers, swallows, sparrows, warblers, and blackbirds. Small landbirds (perching birds, or passerines), com-pose the largest group of birds worldwide, but in the context of this book, they are a relatively small group.

How Coastal Birds Live

The cycles of a bird's life are governed by the seasons. Birds feed, flock, migrate, and nest in response to the rhythm of the seasons and the surrounding environmental conditions.

Because the majority of the species encountered along the southern California coastal strip are waterbirds, the discussion below emphasizes interesting waterbird behaviors. Furthermore, seabirds and shorebirds, species that are consistently associated with coastal locations, receive the most attention.

Feeding and Flocking

Shorebird Feeding

Birding on the open mudflats offers a unique window on the feeding behavior of shorebirds. Tidal mudflats are a giant larder thrown open twice a day, when the ebb tide reveals the wealth of food available to the waiting birds.

When scanning a group of feeding shorebirds, notice the dif-ferences in bill and leg lengths. The variety of bill sizes and shapes indicates they eat an array of food. Differences in leg length re-flect different feeding locations on the marsh or mudflat.

Ornithologists speculate that feeding in different locations on the mudflats may lessen the competition for food sources in a uniform habitat. At an estuary, Semipalmated and Black-bellied

Plovers *(Charadrius semipalmatus* and *Pluvialis squatarola)* prefer dry expanses of hard-packed mud. Least Sandpipers *(Calidris minutilla)* scurry around at the margins of shallow pools, nabbing insects. At deeper pools, Western Sandpipers *(Calidris mauri)* probe at the waterline with their tapered bills. Dunlin *(Calidris alpina)* and Short-billed Dowitchers *(Limnodromus griseus)* feed in still deeper water. Greater Yellowlegs *(Tringa melanoleuca)* prance through the middle of puddles, changing direction to surprise minnows. The tallest waders—the Long-billed Curlews *(Numenius americanus),* Marbled Godwits *(Limosa fedoa),* and Willets *(Catoptrophorus semipalmatus)*—roam everywhere, using their longer legs to walk far out into the deepest channels of the estuary.

Shorebirds choose a variety of prey. Marbled Godwits devour several species of polychaete worms (particularly *Neanthes succinea),* an abundant prey item found in estuaries. On the beach in fall, Sanderlings *(Calidris alba)* probe for immature Western Sand Crabs *(Emerita analoga),* which burrow backward in the wet sand every time a wave recedes. Whimbrels *(Numenius phaeopus)* seek beach hoppers (amphipod crustaceans) that hide under piles of decaying kelp. Snowy Plovers *(Charadrius alexandrinus)* snatch beetles (Coleoptera) and kelp flies (Coleopa) on the dry sand of the upper beach.

The amount of time a shorebird can feed is limited by the tides. Plovers and godwits often feed at night, when more prey is exposed by the low tide. Fortunately, the greatest minus tides in winter occur during the day, a boon to wintering shorebirds along the southern California coast.

Some shorebirds set up small temporary territories around their feeding sources in winter. When a solitary bird finds a good spot that yields sand crabs or worms, it will often chase away trespassers.

Shorebird Flocking

But feeding alone has risks. The flocking behavior of a shorebird is as crucial a mechanism for survival as its feeding pattern. In an open environment with few places to hide, flocking together in groups of varying size helps protect individuals from marauding raptors. One study indicated that isolated shorebirds were three times more likely to be attacked by predators than were flock

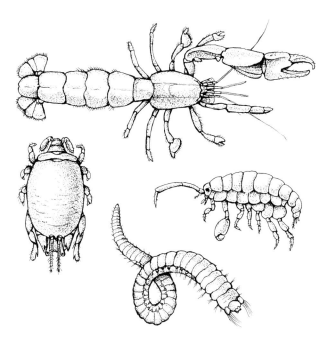

Figure 11. Shorebird invertebrate prey (counterclockwise from top):
California Ghost Shrimp *(Callianassa californiensis)*, Western Sand Crab
(Emerita analoga), polychaete worm *(Amphicteus scaphobranchiata)*,
Beach Hopper *(Orchestoidea corniculata)*.

members. Another study found that the major source of predation for wintering Dunlin was raptors.

Shorebirds apparently benefit by flocking together while
roosting, too. The more birds in the flock, the more watchful eyes
available to alert the group to an approaching predator. If the
flock is too small, however, the birds will be jittery and nervous,
responding to false alarms too often. If the flock is too large, individuals may get careless and be less responsive to the perils of attack. Several studies suggest that the birds in a flock of intermediate size reap the most advantages.

When you see the graceful twisting and turning of shorebird

flocks in flight, a question springs to mind: Are the birds following a leader? What is it that keeps the birds together — an equal
distance from each other yet never colliding, seldom behind,
never in front, but always in unison?

Biologists have described many instances of self-organization
in nature, such as the coordinated flocking of shorebirds and the
schooling of fish. Unlike most human interactions, patterns in
nature often operate without a designated leader. A group works
in a decentralized manner, each member of the unit following instinctual rules of behavior. Thus shorebird flocks rely on the ability of each bird to steer clear of nearby individuals, maintain the
average position of those birds in the lead, and stay an average
distance from their immediate neighbors.

In addition, the loud cries of shorebirds, such as those of the
Black-bellied Plover, Greater Yellowlegs, Willet, and Whimbrel,
serve to warn other shorebirds when danger threatens the flock.
You would assume this would call the predator's attention to the
screaming individuals, but instead it gives the other shorebirds a
chance to gather together immediately into a flock, which the
calling birds can then join.

Seabird Feeding

Like shorebirds, seabirds sort themselves out to avoid direct competition for prey. Loons dive deep beneath the ocean's surface;
terns plunge-dive from above; shearwaters snatch food from the
surface of the water; gulls scavenge at the leftovers.

More importantly, seabird feeding patterns are governed by
the waxing and waning of the productivity of the sea. At certain
seasons and certain locations, offshore waters teem with fish and
plankton. For example, the prevailing northwest winds of spring
promote upwelling in the northern portion of the Southern
California Bight and in areas off Point Conception. Cold water
rich in nutrients is brought up from the bottom of the ocean,
nourishing a burst of the tiniest plants of the ocean: the phytoplankton. Phytoplankton form the base of the food web for all
ocean creatures, so areas of upwelling are rich in fish. Migrating
seabirds in spring, such as Pacific Loons *(Gavia pacifica)*, take
advantage of upwelling zones where fish aggregate. They select
stopover habitats in the northern portion of the Southern California Bight and near Point Conception.

Point Conception generally divides the more northerly bird

species from the more southerly ones; however, southern California's offshore waters contain both northern and southern influences, so we see species along our coast that prefer colder, more northern waters, as well as species that inhabit warmer, more southern waters. Seabirds can be classified by these preferences (table 2), although they may sometimes wander north or south of their normal habitats, depending upon oceanic conditions.

TABLE 2. Species Indicating Warmer or Cooler Waters

WARMER WATER	COOLER or UPWELLING WATER
Black-vented Shearwater	Sooty Shearwater
Heermann's Gull	Western and Clark's Grebes
Brown Pelican	Western Gull
Elegant Tern	California Gull
Royal Tern	Pelagic Cormorant
	Pigeon Guillemot

Source: In part from Briggs et al. 1987, cited in Baird, Patricia Herron 1994, "Birds" (chapter 10). In *Ecology of the Southern California Bight: A Synthesis and Interpretation,* edited by Donald J. Reish and Murray D. Dailey. Berkeley and Los Angeles: University of California Press, p. 561.

The single most important prey item to a number of seabirds is the Northern Anchovy *(Engraulis mordax).* Indeed, so dependent is the Brown Pelican *(Pelecanus occidentalis)* on the anchovy that the bird's breeding success fluctuates in direct proportion to the abundance of the anchovy in any given year. The same can be said for the Elegant Tern *(Sterna elegans).* When anchovies are plentiful, Elegant Tern colonies prosper.

Because the anchovy is such an important food for seabirds, scientists regularly monitor it. During El Niño–Southern Oscillation events, when abnormally warm waters are found off the southern California coast, anchovy populations are affected. During the 1982/83 El Niño, anchovies moved to deeper, cooler water to spawn farther north and farther offshore, resulting in a hardship for many seabirds. In the 1997/98 El Niño, Brown Pelicans suffered a downturn due to the lack of available anchovies to feed their young. In 1999, when more normal environmental conditions returned offshore, Brown Pelican numbers rebounded.

Besides anchovies, seabirds consume mostly squid, euphausids (shrimplike crustaceans called krill), and rockfish.

Seabird Flocking

How do seabirds, which forage alone, find concentrations of prey? Seabirds do not gather in regular feeding flocks outside of the breeding season the way shorebirds do. Except for Double-crested and Brandt's Cormorants *(Phalacrocorax auritus* and *P. penicillatus),* which sometimes cooperate by fishing in groups, seabirds forage individually. They come together temporarily when prey is sighted, then disperse.

Figure 12. Heermann's Gull stealing Brown Pelican catch.

Seabirds have an uncanny way of locating food from great distances. When a mixed-species flock of seabirds comes together in this way, it creates a feeding frenzy. Whether you are scanning from a coastal promontory or from a boat offshore, pay attention when large groups of seabirds gather to feed; it is an exciting spectacle. In fall, feeding groups of seabirds frequently come unusually close to shore in pursuit of dense schools of anchovies. (Dolphins, too, chase the anchovies.)

Gulls discover the schooling fish first. The dark gray forms of Heermann's Gulls *(Larus heermanni)* appear out of nowhere and land on the water. A group of Brown Pelicans circles overhead. Brandt's Cormorants, alighting in twos and threes, add to the fray. The feeding frenzy escalates.

Soon the surface of the water churns with fish. The Brown Pelicans plunge-dive from above, scoop the anchovies into their throat pouches, and gulp them down. If the Heermann's Gulls can steal the pelicans' catch they will. A phalanx of Sooty Shearwaters *(Puffinus griseus)* glides by and plops down into the water. Western Gulls *(Larus occidentalis),* lured by the commotion, descend to take part. They join the throng of birds grabbing and bobbing on the surface or diving underwater to chase down the fish. Slurping and wrestling with the anchovies, the shearwaters and gulls are the most aggressive. The long-necked cormorants, heads held erect like models of dignity in the midst of the confusion, dive sedately when they can find an opening in the crowd. Elegant Terns hang on the fringes of the madness, diving when they find a target. The cries of the gulls and terns carry far across the ocean.

As the school of anchovies moves through the water, trying to avoid the crazy mob of birds, the birds move with it. Diving, plunging, snatching from the surface, or simply scavenging the leftovers, the seabirds feed from the writhing swarm of fish. At the peak of the frenzy, you can see hundreds of birds of several species — an amazing display of seabird feeding behavior (see fig. 13).

Migration

Along the southern California coast, we have a front-row seat at the spectacle of bird migration. A broad north-south route, called the Pacific Flyway, includes a corridor following the California coastline. The exciting times are spring and fall, when birds are on the move. In spring, many species fly north from southern wintering grounds in Mexico and Central and South America. In summer and fall, they return, some from Arctic breeding territories, some from the Northwest, some from the Great Basin. Often they hug the coastline on their journeys, giving us a window of opportunity to study them as they come and go.

Seabird Migration

One of the most wonderful aspects of southern California birding is the spring seabird migration. From March through early June, but primarily in April and early May, thousands of loons, cormorants, brant, scoters, phalaropes, gulls, and terns migrate past our coast to breeding grounds in Canada and Alaska.

The configuration of the southern California coast is an im-

Figure 13. Seabird feeding frenzy offshore.

portant factor in seabird migration. Migrating seabirds on their way north move parallel to the coast, which generally runs in a north-south direction. The stretch of coastline from the Palos Verdes Peninsula to Point Conception, however, has an east-west orientation, which acts as a barrier to northward-flying seabirds. Thus, migrating birds veer to the west, hugging this portion of the coastline until they round Point Conception, when they are once again able to follow a more northerly direction. Along this east-west portion of the coast, migrating seabirds often pass very close to shore, and good seabird watching occurs from promontories such as Point Vicente, Point Dume, Mugu Rock, and Goleta Point.

North and south of this east-west portion of the coast, seabirds usually travel farther offshore; however, strong afternoon westerly winds — often exceeding 20 miles per hour — may push the birds closer to shore, affording good seabird watching off La Jolla or at Newport Pier to the south, and Montana de Oro State Park and Point Piedras Blancas to the north.

Pacific Loons are the most numerous species in spring migration, their passage peaking in mid- to late April. They winter along both coasts of Baja California and in the waters of the Gulf of California, then mass in staging (gathering) areas on the west coast of Baja to molt into spring plumage and fuel up for the

northward journey. By watching from coastal promontories, observers can tabulate almost the entire population of Pacific Loons that have wintered to the south. In spring of 1996, a carefully monitored count of Pacific Loons passing Point Piedras Blancas calculated 456,333 individuals migrating that season.

Watching the spring seabird migration from shore can prove challenging on a windy day, but the sight of masses of seabirds passing north is worth it. The fierce west wind makes it difficult to keep a spotting scope steady. The sea is a roiling, heaving collection of whitecaps and green, choppy waves. Gulls careen across the sky. Carried helter-skelter, they plunge and dip, rise and fall, making slow progress against the ferocious wind. Most are California and Bonaparte's Gulls *(Larus californicus* and *L. philadelphia)*, difficult to discern against a background of rough seas. Forster's and Royal Terns *(Sterna forsteri* and *S. maxima)* hover and wheel among the gulls.

Through the chaos of the winds blowing and the gulls flying, watch for the Pacific Loons. On windy days, the loons pass low over the water's surface in groups of 10 to 30. They fly in loose, rather straggling flocks. As they fly by, their black throat patches are visible first, then the soft gray of their napes, then the white speckling on their dark wings. Sometimes the loons come so close you can see their necks pumping up and down with the force of each strong wingbeat. Heads bowed forward, necks flexing with the effort of battling the wind, feet held straight out behind, they fly on relentlessly.

Flocks of Brant *(Branta bernicla),* small black-and-white sea geese, appear as undulating skeins on the horizon. Groups of 100 to 200 individuals are interspersed with large flocks of up to a thousand or more. Along with Pacific Loons and Brant, Surf Scoters *(Melanitta perspicillata)* are abundant migrants. The scoters fly fast in straighter lines than the Brant. From a distance, they are little black dots massing above the ocean surface. As they approach, you can see their wings beating like whirring propellers as they hurl themselves northward. The males can be recognized from afar by their bright red bills and white forehead and nape patches.

The seabirds' need to push northward, whatever the obstacles, lends an air of urgency and excitement to spring seabird watching. Although the variety of species is not large, the numbers of individuals moving past is huge. Once you learn to recognize seabirds' shapes and behaviors, tune in to one of the greatest migration shows in North America (see fig. 15).

Shorebird Migration

Long before the foggy mornings of summer give way to the clear ones of September, the fall shorebird migration is underway. Fall shorebird migration commences in June, as the first of the shorebirds nesting in northern Alaska begin to trickle southward from their breeding grounds. It is more prolonged than spring migration, allowing us time to study the birds as they come through. The adult males depart first, followed by the females. By early August, the juvenile birds have begun to appear along our coast.

Come July and August, southern California's mudflats and estuaries are inundated by flock after flock of "the wind birds," as author Peter Matthiessen calls them in his book by that name (1999). Blowing southward from Alaska to Baja California and even to the tip of Argentina, the shorebirds chart their way, searching for inlets, estuaries, and river mouths where they can rest and revive en route.

Some of these shorebirds will remain to spend the winter in southern California at the larger wetlands from Morro Bay south to San Diego Bay. But it is during the shorebird fall (late June through September) that the largest and most varied assortment

Figure 14. Seawatch.

of these delicate, long-winged migrants visits our shores before resuming the southbound journey.

Shorebirds depend upon a series of intertidal wetlands for stopovers along their migration route, a prime reason to protect and maintain a chain of wetlands along the coastal corridor. According to one shorebird survey, the most heavily used locations along the southern California coast are San Diego Bay, the San Diego River Flood Control Channel, Bolsa Chica, Mugu Lagoon, and Morro Bay.

At the height of fall shorebird migration, in August, Western Sandpipers are the most numerous species along the coast, crowding the mudflats, feeding incessantly. With the advent of satellite telemetry, by which individual birds can be tracked by radio transmitters during their migrations, scientists have been able to learn their habits. We know Western Sandpipers put on extra fat and muscle for fueling the long flights to and from breeding grounds. We know they have a tendency to return to the same wintering grounds year after year. Called site fidelity, this behavior maximizes species survival. Furthermore, we know that male Western Sandpipers winter farther north along the coast than females (similar to several other shorebird species). Presumably this is so the males can get a head start in spring to reach

Figure 15. Seabird migration along the coast.

their Arctic breeding grounds, where they must arrive early enough to secure a good territorial patch in order to attract the females, who show up several days later.

Waterfowl Migration

For years it was assumed that most ducks and geese migrated pretty much within one of the four north-south flyways across North America: Atlantic, Mississippi, Central, and Pacific. More recently, banding experiments show that ducks do not always move in a north-south direction, and that these flyways are more like broad corridors with blurred borders.

Depending on the severity of the weather to the north, most waterfowl that use the Pacific Flyway spend the winter in the Sacramento and San Joaquin Valleys in northern California. Aside from the scoters, which are coastal migrants, many ducks travel the Pacific Flyway farther inland, coming to rest in freshwater reservoirs and lakes away from the coast. Still, coastal southern California attracts its fair share of migrant and wintering diving and dabbling ducks. The ducks arrive by December, and most are on their return trip north by March.

One of the paradoxes of southern California birding is that good bird-watching opportunities often lie in the midst of the

hustle and bustle of an urban environment. For example, the mouth of the San Diego River Flood Control Channel near Mission Bay provides excellent waterfowl watching—the background of San Diego's tall buildings and busy freeways only heightens the contrast to a river channel packed with an array of brilliantly plumaged ducks.

As the sun dips low in the western sky on a clear winter evening, conditions are perfect for viewing the assorted species. In the shallow waters, Blue-winged, Cinnamon, and Green-winged Teal *(Anas discors, A. cyanoptera,* and *A. crecca)* poke about on the fringes of the weedy grasses. The rich chestnut of the male Cinnamon Teal catches your eye first. One of our earliest migrants, this pair of Cinnamon Teal is probably already on its journey north, although it is only January. Gadwalls *(Anas strepera),* Northern Shovelers *(A. clypeata),* and Northern Pintails *(A. acuta)* paddle to and fro. In full nuptial plumage, the males display the green, rust, tan, and black of their beautiful feathers.

A striking male Hooded Merganser *(Lophodytes cucullatus)* stands on the sidelines preening. With his crest raised, he is a fine combination of black and white with rusty sides. Among the hundreds of American Wigeons *(Anas americana),* the rare Eurasian Wigeon *(A. penelope)* can occasionally be found, its rufous (reddish) head with golden crown shining in the evening light.

In the deeper water toward the mouth of the river, Lesser Scaup *(Aythya affinis)*, Buffleheads *(Bucephala albeola)*, and Ruddy Ducks *(Oxyura jamaicensis)* swim and dive.

And all this bird activity can be viewed from a convenient bike path adjacent to the river channel, making it an ideal place for a quick respite from the fast pace of modern life.

Other Migrations

Other bird species travel up and down southern California's coast, migrating in response to climate and food conditions. Many of our local raptor populations are augmented by birds coming from inland or northern breeding grounds to spend the winter.

Several species of seabirds perform a "reverse migration," migrating *south* in late winter and spring, contrary to the migration path of most birds, which fly *north* in spring. For example, Black-vented Shearwaters *(Puffinus opisthomelas)*, Brown Pelicans, Heermann's Gulls, and most Royal Terns fly *south* to breed off Baja California and western Mexico. After breeding, they return *north* in late summer and fall.

Yet another migration involves the postbreeding movements of herons, ibises, terns, and some gulls. In an attempt to utilize food sources away from their original nesting colonies, many of which are overcrowded, certain species practice what is called postbreeding dispersal. This term applies to the movements of adult and juvenile birds in various directions away from the breeding area. The dispersal can be to the north or south and is temporary. In late summer and fall, the young—occasionally accompanied by the adults—of White-faced Ibis *(Plegadis chihi)*, Elegant Tern, Black Skimmer *(Rynchops niger)*, and a variety of herons often move around after nesting. Some move north along the coast from nesting colonies near Bolsa Chica, San Diego, or Baja California. Immature birds of certain heron species, such as the Tricolored Heron *(Egretta tricolor)* and the Reddish Egret *(E. rufescens)*, have shown up at areas along the southern California coast in fall as part of postbreeding dispersal. These are rare birds, termed vagrants, which have wandered outside their normal ranges.

With the onset of winter, birds that indulge in postbreeding dispersal usually return to their traditional ranges and feeding grounds. Or, if they find ideal conditions during postbreeding dispersal, they sometimes decide to spend the winter or even remain to nest.

Nesting Colonies

Several species of waterbirds that breed along the southern California coast nest in colonies: Brown Pelicans, Double-crested, Brandt's, and Pelagic Cormorants *(Phalacrocorax pelagicus)*, Western Gulls, Royal, Elegant, Forster's, Least, and Caspian Terns *(Sterna caspia)*, Black Skimmers, and Pigeon Guillemots *(Cepphus columba)*.

Brown Pelicans, the three cormorants, Western Gulls, and Pigeon Guillemots require nesting sites near their food sources in ocean waters. They seek isolated bluffs, sea cliffs, and islands, where they use every bit of space. Most nesting colonies are on offshore islands or along the rocky mainland coast north of Point Conception. On the Northern Channel Islands, the Brandt's and Double-crested Cormorants choose the flatter surfaces, the Pelagic Cormorants take the steepest ledges, and Pigeon Guillemots nest in crevices near the water's edge.

Terns and skimmers, on the other hand, require flat sandbars for nesting, preferably protected by a lagoon or surrounded by water. They frequently settle on human-made islands or dikes that have been provided for them. Colonial nesters on sandy beaches, such as California Least Terns *(Sterna antillarum browni)*, are more vulnerable to high tides, late storms, and predation than those on remote rock stacks or steep seacliffs.

American Avocets *(Recurvirostra americana)* and Black-necked Stilts *(Himantopus mexicanus)* nest in loose colonies on the sandy margins of estuaries and marshes, such as at San Joaquin Wildlife Sanctuary and Upper Newport Bay. Snowy Plovers nest in groups on sandy beaches.

Heron species are gregarious, too. Great Blue Herons *(Ardea herodias)*, Great Egrets *(A. alba)*, Snowy Egrets *(Egretta thula)*, and Black-crowned Night-Herons *(Nycticorax nycticorax)* place their nests close together in tall trees near the coast. They customarily choose groves of eucalyptus or cypress adjacent to lagoons or sheltered harbors so as to be near a handy food source.

In addition, several species of landbirds found near the coast nest colonially, notably Cliff Swallows *(Petrochelidon pyrrhonota)* and Red-winged Blackbirds *(Agelaius phoeniceus)*. Barn and Tree Swallows *(Hirundo rustica* and *Tachycineta bicolor)*, European Starlings *(Sturnus vulgaris)*, and Brewer's Blackbirds *(Euphagus cyanocephalus)* also form breeding colonies, but in smaller numbers.

Monitoring Seabird Colonies

Because seabird colonies are highly visible and often occupy the same sites year after year, measuring their success or failure helps biologists understand changes in fish populations, climatic conditions, and the presence or absence of environmental toxins.

At the South San Diego Bay Saltworks (a unit of the San Diego National Wildlife Refuge), nesting success of multispecies seabird colonies has been monitored regularly for many years. Depending on the status of the fish prey base, particularly anchovies for the Elegant Terns, and depending on the disturbance and loss to predators (coyotes and Peregrine Falcons *[Falco peregrinus]*), the number of breeding pairs fluctuates from year to year.

For example, in 1999 at the South San Diego Bay Saltworks, there were 84 Double-crested Cormorant nests, 36 Royal Tern nests, and 3,100 Elegant Tern nests. It was a very good year for anchovies. In 2002, much lower nesting success was suggested by preliminary totals: 49 nests for Double-crested Cormorants, 1 to 3 nests for Royal Terns, and 37 to 100 nests for Elegant Terns. In that year, a lack of local feeder fish was reflected in reduced adult numbers, reduced nesting attempts, reduced clutch sizes, and increased chick mortality.

Moreover, the expansion of the Elegant Tern's range into southern California is a good illustration of nesting success in a food-rich year. Formerly, Elegant Terns were unknown as nesters in San Diego County and were considered rare in California. But in 1959, several Elegant Tern pairs appeared and nested for the first time at South San Diego Bay. It is thought this first nesting attempt was tied to a sudden increase in anchovy abundance in waters off northern Baja California and southern California, perhaps as a result of the La Niña (colder water) conditions following the 1957/58 El Niño. Because this outpost was successful, Elegant Terns have returned regularly to nest in colonies in San Diego, Orange, and Los Angeles Counties, where their population is generally increasing.

Examples of Nesting Colonies on the Southern California Coast

At Morro Bay State Park, a busy, noisy colony of herons, egrets, and cormorants nests in the eucalyptus and cypress trees. From February through May, the stick nests hold approximately 20 pairs of Great Blue Herons, 35 pairs of Great Egrets, 250 pairs of Double-crested Cormorants, and several pairs each of Black-crowned Night-Herons and Snowy Egrets.

Constantly coming and going, the Great Blue Herons and Great Egrets flap slowly in to land on the rims of their nests. The young Great Blue Herons, which lie quietly in the nest between feedings, become excited when the adult arrives. The awkward shapes of the nestlings can be seen bobbing and wobbling within the nest, awaiting a meal of regurgitated fish. When the adult feeds the youngster, it lowers its head in a pumping motion. The young heron grabs the proffered bill crosswise. In what looks more like a wrestling match than a feeding session, food is given to each of the young herons. Often, the pushiest nestling gets the most nourishment.

At Sands Beach at Coal Oil Point Reserve, a sizeable colony of Snowy Plovers has reestablished itself at a former nest site that had been abandoned for decades. Now protected by a low rope fence and monitored by a series of docent naturalists, the Snowy Plover colony is beginning to expand. In 2001, one chick fledged, in 2002, 14 chicks fledged, and in 2003, 39 chicks fledged. In 2004, however, only 27 chicks fledged, due to predation by crows, skunks, and Red-tailed Hawks *(Buteo jamaicensis)*, as well as some unusually high tides.

On the dry sand littered with bits of shells and small stones, the Snowy Plovers remain almost invisible on their nests. They

Figure 16. Western Snowy Plover.

space themselves several feet apart to be less obvious. For the nest, the male makes a slight depression in the sand and lines it with tiny pieces of driftwood or other beach debris. Males and females take turns incubating the eggs: males by day, females by night. When the adult leaves the nest, it runs swiftly in hunched posture, so that its darker markings are hidden and only the pale back shows. After running a suitable distance from the nest, the bird stands upright again to begin its search for food.

A Snowy Plover nestling looks like two balls of fluff—one for the head and the other for the body—that have sprouted legs. The chick's down is marked with dark squiggles. When the male Snowy, who watches the chicks until after they are fledged, senses danger he signals to the young ones. Immediately, they squat in the sand. Camouflaged like this, the young plovers escape the notice of most predators. The greatest danger comes from human beachgoers inadvertently stepping on the motionless chicks.

On a dredge spoil island at the Port of Los Angeles, 5,500 pairs of Elegant Terns nest shoulder to shoulder (an arrangement known as hexagonal packing) in a seething, screeching mass of crested black-and-white. Each tern nest contains a single egg lying on the flat surface, guarded by two adults: while one sits on the egg to protect it from the sun, the other stands guard. Orange bills pointed skyward, the adults protest loudly if their tiny space is violated by an adjacent couple. The noise is deafening.

In July at Bolsa Chica Ecological Reserve, you can visit a famous tern nesting area in southern California. Standing on the boardwalk that separates the two *bolsas* (Spanish for "purses"), or lagoons, of the salt marsh, look for Caspian, Elegant, and Least Terns, as well as Black Skimmers. The birds are ferrying fish back and forth to their nesting colonies on two sandbar islands to the south. The white forms of the terns, their varied cries, and the fish wiggling in their bills all set the stage for a good morning of birding.

If you take your spotting scope and walk southward along the coastal highway, you arrive at a place that has a good view of the tern and skimmer colony nesting on the North Island. A group of Black Skimmers nests at the far end of the island, Elegant Terns nest in the middle, and a cluster of Caspian Terns claims the near end.

Groups of young terns are known as "creches" and serve as a gathering place for the fledglings while they are being fed by the

Figure 17. Tern and Skimmer nesting colony.

adults. Each tern species has young in various stages of development. The juvenile Elegant Terns are the most numerous, begging assiduously from the adults. When an Elegant Tern alights, bedlam ensues as several chicks fight for the fish flapping in the adult's orange bill. After the most aggressive chick gets its meal, the adult is nearly brought to the ground. The downy Caspian Tern chicks are not as mature as the Elegants, but they are already much bulkier. There are fewer Caspian chicks, and they keep to themselves.

The young Black Skimmers steal the show. Looking through the scope, at first you see only the adults nestled in the sand. Finally, when one of the skimmers stands up, a chick appears from underneath. The little thing looks surprisingly alert for having been sat on all that time. A skimmer chick resembles an adult, complete with thick mandibles. The chick's upper and lower mandibles are, however, more nearly the same length than the adult's, allowing it to grasp squirming fish brought by the parent. Only when the chicks are old enough to learn to skim does the lower mandible begin to grow longer than the upper.

Skimmer chicks are good at digging. They run to an adult and stand between its two legs, then lie prostrate in the sand and scratch backward with one foot then another to make a little depression. Once the hollow is dug, the chick lies flush with the sand, bill outstretched. In a minute, the adult will sit down on top of the half-buried chick to insulate it from the heat of the sun. A chick unshielded from the sun would quickly expire.

With the highway buzzing behind you, and the terns and skimmers flying in and out of the nesting colony in front of you, this is the epitome of southern California birding. Most human commuters, intent on their busy schedules, know nothing of the busy schedules the tern and skimmer parents must meet that morning as they commute to and from the ocean with food for their chicks.

Advantages and Disadvantages of Colonial Nesting

Why do birds nest in colonies? Studies show that colonial nesting has two benefits for birds: exchange of food-getting information, and, most of the time, reduced danger of predation.

Seabirds nest together and forage together during the breeding season, because their food supplies are unpredictable and may change from day to day. Thus, in most species, the more experienced members of the nesting colony serve as guides to food sources for the younger or less-knowledgeable birds, who follow them out to hunt. But some species that nest in colonies, such as Great Blue Herons, hunt alone. In their case, perhaps the benefit of a colony to the older, more experienced birds is protection from enemies. Presumably, the older birds retain the best nest locations at the center of the colony, shielded from predators. This selfish herd theory postulates that the optimal location for a nest would be at the center of the colony.

Furthermore, nesting colonies often comprise more than one species. At several sites in San Diego County, Snowy Plovers that nested within Least Tern colonies gained protection from predators such as Common Ravens *(Corvus corax)*.

Colonial nesting can have disadvantages, however. Ornithologists monitoring seabird colonies found that large, loose colonies are more vulnerable to aerial predation than dense, small colonies. Similarly, scientists found that large Least Tern or Snowy Plover colonies were more vulnerable to destruction than smaller ones. Because of human encroachment on their nesting habitat, Least Tern and/or Snowy Plover colonies are concentrated at a few sites on the sandy beach. These stable, large colonies tend to attract more predators. Thus, unnaturally large colonies put terns and plovers at the mercy of a predator or a high tide that may result in unnaturally large losses.

Before You Go

Birding Equipment

One of the delights of birding is the relative lack of equipment needed. Birders grab a pair of binoculars and a field guide and are set to go. Still, consider carefully which binoculars and field guide to purchase.

Binoculars

For birders and naturalists, binoculars are indispensable tools to explore the natural world. Birders seek to identify and name the birds they see. In order to do this, a birder must look through binoculars, which magnify the bird so that all its field marks are visible. Magnification, or power, means the number of times the object viewed is larger than when seen with the naked eye. Figures such as "7 × 35" stamped on the binoculars refer to the magnification (the first number) and the diameter of the objective lens (the second number). Examples of common power/size combinations are 7 × 35, 8 × 42, and 10 × 42.

In discussing the relative merits of different pairs of binoculars, an understanding of magnification, brilliance, resolution, field of view, close focus, and eye relief is necessary. Other considerations are weight and the feel of the binoculars. Never before

have birders had such a wide choice in the models and prices. With the growing numbers of birders, manufacturers have hastened to produce an array of binoculars suited to every birder's needs, whether beginner or expert. The rule of thumb here is to buy the best binoculars you can afford.

Spotting Scopes

Spotting scopes are an optional addition to a birder's standard equipment. Looking at faraway birds makes a spotting scope highly desirable, if not a downright necessity. A scope brings birds from 15 to 60 times closer, much closer than even the most powerful binoculars.

Field Guides

The other essential element of a birder's gear is the field guide. It is a book, preferably one light enough to carry with you into the field, filled with photographs or artists' plates depicting birds. A birder can never have too many field guides. And, as with binoculars, there have never been so many good ones to choose from. There are field guides for novices and for advanced birders. There are field guides for regions of North America and for all of North America. There are field guides for certain groups of birds, such as warblers or sparrows.

Most current field guides are excellent. Whether you choose a field guide organized taxonomically or in some other way, whether it is illustrated with artwork or photographs, whether it has range maps opposite the bird illustrations or not, choose the field guide that works best for you.

Along with a field guide, a checklist of the birds of your area is helpful. The best checklists are those that designate abundance (common, uncommon) and seasonal distribution (fall, winter) so you can gauge the chances of seeing a particular species at a given time of year.

This book might serve as a field guide and a checklist for the common birds of the southern California coastal strip.

Travel Tips

Before exploring southern California's birding locations, here are some suggestions.

Traffic is a major consideration. If you set out for birding sites in Los Angeles, Orange, or San Diego Counties, avoid peak rush hours on weekdays. Also, parking at the beaches on summer (June through August) weekends can be difficult. Arrive as early as possible.

The state of the tides (high or low) is important. Study the tide tables, available free at bait shops. A pleasant stroll on the beach can turn into a rock-hopping challenge if you are stranded by high water. The best time to bird most areas is at midtide, that is, neither at high tide nor at low tide but somewhere in between. A couple of hours toward the first low tide of the day is best, especially for watching shorebirds as they feed on mudflats. When the tide is very low, the birds, which usually feed at the water's edge, will be too far away. For example, if you're going to Morro Bay to see American White Pelicans *(Pelecanus erythrorhynchos)* and Brant *(Branta bernicla),* time your visit to coincide with at least one high tide so the bay will be full. To see American Avocets *(Recurvirostra americana),* Long-billed Curlews *(Numenius americanus)*, or Marbled Godwits *(Limosa fedoa)*, be there when the tide ebbs and mudflats are exposed.

Lighting is another help or hindrance, depending upon the situation. For morning seabird watching, have the light at your back. Looking at any kind of bird while you squint through binoculars against the glare of afternoon sun is a waste of time.

For watching seabirds, make use of the cliffs and bluff tops to look out over the ocean. If you take up a position above the waterline at convenient overlooks, you can see farther offshore. And, for looking through a spotting scope, the higher you are, the less heat haze distortion — besides, birds flying by may be at eye level.

Coastal birds are active throughout the day, but morning viewing is usually the most rewarding. The greatest movement of seabirds in spring migration takes place in early morning, just after dawn. There is plenty of activity at other times of the day, but the numbers may not be as great. Wind can be a factor in the afternoons, sometimes blowing the birds close in, and at others keeping them too far offshore.

Studying shorebirds is different from watching seabirds. With shorebirds, get as close as you can to the birds. If you move slowly, shorebirds — intent upon feeding — will allow close approach. If you seek landbirds, early morning viewing is essential. The earlier you start searching for the bird species of coastal sage scrub, the more you will see and hear.

Last, but perhaps most important of all, get good maps of the areas you intend to bird. Those available from the Automobile Club of Southern California (AAA), as well as the *Thomas Guide* and the *DeLorme Atlas,* are excellent. If you prefer to work with a computer, many good mapping programs are available. This book has maps and written directions, but you may need regional maps to get an overall view of the area and to choose the quickest, most direct routes from wherever you are staying.

A few cautionary words about birding sites in urban areas: always lock your car, do not leave valuables visible in your car, and take your optics with you.

The following species accounts provide a representative sample of birds that might be observed on the southern California coast. All of the birds that occur in the area cannot be included. In each account, the facts presented vary according to which aspects of the species are deemed important. Note that additional information about many of the species is found in "How Coastal Birds Live."

At the end of each account, a brief description of the bird's plumage is further classified as breeding or nonbreeding, and male or female, if those categories apply. For example, if the bird's nonbreeding and breeding plumage are similar, or if the male and female plumage are similar, all of the categories are not included. Immature plumages (including, in some cases, juvenile plumages) are described if they differ markedly from the adults' nonbreeding plumage. In some species, a description of both breeding and nonbreeding plumages is omitted, if only one plumage is likely to be obeserved in southern California (e.g., some adult gulls depart our region before they attain their breeding plumage).

Vocalizations are mentioned if they are likely to be heard or noticed by an observer.

For definitions of seasonal status, abundance, breeding, and protection used in the species accounts, see table 3.

GEESE and DUCKS (Anatidae)

The geese and ducks (family Anatidae), called waterfowl as a group, are a large and varied family of birds found all over the world. Many of these species are remarkable for their large, gregarious flocks and spectacular long-distance migratory movements.

All waterfowl are excellent swimmers. Their three front toes are webbed; the fourth toe is smaller and placed slightly higher on the foot than the others. When swimming, they bring the foot forward through the water by folding the toes and webs together, then on the backward stroke, the webs are fully spread to push the bird forward in the water.

The legs of geese and ducks are positioned on their bodies in

TABLE 3. Definitions of Seasonal Status, Abundance, Breeding, and Protection

STATUS	
Resident	Known or presumed to breed in our area. Individuals may be either year-round residents or summer residents, which appear in spring and stay through the summer.
Visitor	Does not breed in our area. Most individuals are winter visitors, but some can be found here in other seasons or even year-round.
Migrant	Passes through our area in spring en route to breeding grounds or in fall en route to wintering grounds.

ABUNDANCE, BREEDING, and PROTECTION	
Common	Almost always encountered in proper habitat in moderate numbers.
Fairly Common	Usually encountered in proper habitat, generally not in large numbers.
Uncommon	Occurs in small numbers.
Rare	Occurs regularly, but generally in very small numbers.
Local	Occurs in certain localized areas.
B	A regular breeder.
B*	An irregular, rare, or very local breeder.
E	Endangered, i.e., in danger of extinction throughout all or a significant portion of its range.
FE	Listed as Federally Endangered.
SE	Listed by the state of California as endangered.
T	Threatened, i.e., likely to become an endangered species in the foreseeable future throughout all or a significant portion of its range.
FT	Listed as Federally Threatened.
ST	Listed by the state of California as threatened.

SEASONS	
Spring	March 1 through May 31.
Summer	June 1 through July 31.
Fall	August 1 through November 30.
Winter	December 1 through February 28.

Source: Abundance definitions in part after Garrett and Dunn 1981.

accordance with their eating habits. Depending upon whether they are adapted for grazing, dabbling, or diving, their legs are placed at the front, middle, or rear of the body. Geese, being grazers, have the legs forward for better balance on land. The legs of dabbling ducks, which feed in shallow water, are placed farther back, making them waddle when walking on land, but graceful when swimming. The diving ducks' legs are at the rear of their bodies; almost helpless on land, they are great divers.

Waterfowl have rounded, flattened bills with a small hook or "nail" at the end. Around the inside edge of their bills, a toothlike fringe acts as a built-in sieve. When feeding, the strong tongue presses against the palate, squeezing excess water and mud out of the bill, leaving the food inside.

Like many other species of waterbirds, most of the geese and ducks that we see use southern California as a wintering destination. They have nested in areas to the north and arrive in fall and early winter when water on their breeding grounds freezes and food becomes scarce. In late winter and spring, they will depart again.

Due to the great variety within the waterfowl family, it is divided below into several subfamilies, which are further divided into "tribes."

Geese (Anserinae, Anserini)

Geese (subfamily Anserinae, tribe Anserini) are intermediate in size between swans and ducks. They have long necks and large bodies.

A flock of geese is a highly social organization. A definite pecking order exists: families at the top, mated pairs next, then single adults or yearlings, and immatures at the bottom. Geese usually mate for life, but not always. Regardless, groups of extended families appear to stick together throughout migration and on the wintering grounds.

Geese are exceedingly wary. When a flock of geese graze in a field, one or two individuals act as sentinels and give loud warning calls when danger threatens. Indeed, domesticated geese are occasionally kept as pets in lieu of an alarm system.

Male and female geese look alike; however, the male goose

(gander) tends to be larger than the female goose (technically, she is the goose, although the word "goose" is commonly applied to both sexes).

Two kinds of geese are found regularly along the southern California coast: the Canada Goose *(Branta canadensis)* and the Brant *(B. bernicla)*. The Greater White-fronted Goose *(Anser albifrons)*, Snow Goose *(Chen caerulescens)*, and Ross's Goose *(C. rossii)* pass through irregularly and occasionally spend the winter.

CANADA GOOSE
(Pl. 1)

Branta canadensis

Plate 1. Canada Goose: adult.

In pastures in December or January, a loose flock of geese walks slowly among the stubble. As the geese feed, they bend their long necks to nibble at the waste grain. Although by day they feed in fields, in the evening they will return to roost on coastal sloughs or lakes. They are big, brownish birds with black necks and a large white patch on each cheek that meets under the chin. If the flock is startled, the birds launch into the air with slow, deep wing beats, uttering a chorus of ringing "ka-ronk, ka-ronk" calls. Once aloft, the geese effortlessly place themselves in typical V formation with one bird in the lead. Their flight is surprisingly swift and direct for such heavy birds.

These are Canada Geese, an increasingly common sight in southern California. During the last decade, this species has begun to establish a few year-round, nonmigratory flocks in parks and on golf courses in our region.

Most of our wintering Canada Geese, however, are just visitors having bred far to the north in Canada and Alaska. Others breed from northeastern California to Washington, Oregon, and western Montana. Their close family ties and tendency to nest in the same geographical areas have created isolated populations. As a result, nearly a dozen distinct subspecies have formed, many of which can be identified in the field. They range in size from small-bodied (some about the size of a Mallard [*Anas platyrhynchos*]), up 45 inches tall. Recently, on the basis of DNA analysis, the Canada Goose population has been split into two distinct species: the Cackling Goose (*Branta hutchinsii*), comprising the smaller-sized subspecies, and the Canada Goose *(B. canadensis),* made up of the larger-bodied subspecies. The majority of our flocks are Canada Geese.

SIZE: Length 36–45 in. **ADULT:** Sexes similar. Black head and neck with large white cheek patches meeting under the chin. White belly and flanks. Black bill. **STATUS:** Common to fairly common, but local fall migrant and winter visitor. Uncommon local resident. **(B*)**

BRANT   *Branta bernicla*
(Pl. 2)

At coastal lookouts from La Jolla to Point Piedras Blancas, one of the most impressive migrations of any southern California species takes place: the northward movement of Brant in spring. Flock after flock of this elegant, small sea goose beat their way north from their wintering grounds in Baja California and western Mexico en route to nesting areas in the high Arctic.

Usually in groups of 50 to 75—but sometimes by the hundreds—Brant can be spotted flying up the coast. Their flight pattern is different from that of other geese: they do not use a strict V formation. They string out in long even lines, then bunch up in a crowded cluster. They arrange themselves in a straggling V, then merge together again. When the lead goose flies lower, the whole flock follows; when another climbs, the rest of the birds rise to

Plate 2. Brant: adult.

meet it. Often, the whole flock undulates, flowing up and down like a wave.

Sometimes the Brant flocks are flying far on the horizon, like a waving skein of dark specks. Sometimes they fly just above the sea and close to shore. A birder watching from the bluffs can make out their dark heads, backs, and bellies, and the distinctive white collar around their necks; their mellow honking sounds often carry on the sea breeze.

During migration from late February through early May, Brant stop at coastal estuaries and bays, where they seek out the straplike leaves of eel-grass (*Zostera marina*), which grows in shallow marine waters. At Izembek Lagoon in Alaska, a major staging (gathering) area for Brant in both spring and fall, nearly the entire Pacific Flyway population of Brant congregate to devour the eelgrass meadows growing there.

In fall, Brant migration is completely different. Ornithologists studying the species have established that the birds fly a nonstop trip from Alaska to their wintering grounds in Baja California— a distance of 3,000 miles—in about 60 hours. This puts their flight speed at an amazing 50 miles per hour, and they are seldom visible from shore. (See also page 40.)

SIZE: Length 25 in. **ADULT:** Sexes similar. Small, chunky goose with a dark head, neck, breast, and back. White collar around neck. White underbelly and undertail coverts and white margin of tail. Black bill. **STATUS:** Common spring migrant. Fairly common fall migrant and winter visitor at South San Diego Bay, San Diego River Flood Control Channel, and Morro Bay. Uncommon fall and winter visitor elsewhere.

True Ducks (Anatinae)

The true ducks (subfamily Anatinae), as they are called to differentiate them from the whistling-ducks, can be sorted into four general groups: dabbling ducks (tribe Anatini), diving ducks (tribe Aythyini), sea ducks and mergansers (tribe Mergini), and stiff-tailed ducks (tribe Oxyurini).

In all ducks, the males (drakes) have bright, colorful feathers in contrast to the females' (hens) drab brown and gray outfits. These male/female plumage differences—an example of sexual dimorphism—are an adaptation that allows the dull female to remain camouflaged from predators during nesting.

Nearly all species of ducks also undergo a period called "eclipse plumage," when they molt their breeding feathers and assume a dull nonbreeding plumage, similar to that of the females. Eclipse plumage is like winter or nonbreeding plumage in other bird species, but ducks go through it in the summer. By late fall or early winter, the males have grown their colorful nuptial feathers once again.

Fortunately for southern California birders, many of the ducks do not arrive in our area until November, after they have finished their eclipse period. Even so, identification of individuals in early fall, when the males are still looking nondescript, can be tricky.

Dabbling Ducks (Anatini)

Dabbling ducks are the common surface-feeding ducks—"puddle ducks"—that most of us notice on shallow park ponds, in flooded fields, or at coastal sloughs. Members of this tribe nibble at surface vegetation or upend their bodies with their tails in the air to reach deeper food. Their bills are adapted for straining small invertebrates out of the water or picking up seeds from agricultural fields.

Most male and female dabblers have a colorful wing patch on their secondary flight feathers called a speculum.

Dabbling ducks are suited for living on small ponds surrounded by reeds or cattails. Where the surface area of water is restricted, they have no problem taking off or landing instantly. Unlike the diving ducks, dabblers do not need a running start on open water to become airborne. They jump upward into the air and are off and away.

In short, dabbling ducks are adapted for breeding in areas where a network of small ponds exists, such as in the marshlands of central and northern California, the upper Midwest, and the prairie provinces of Canada. Although a few species of dabblers, stay to nest in small numbers in coastal southern California in summer, our region is generally too dry to support many dabbling ducks year-round.

In southern California, the common species of dabbling ducks are the Gadwall *(Anas strepera)*, American Wigeon *(A. americana)*, Mallard *(A. platyrhynchos)*, Blue-winged, Cinnamon, and Green-winged Teal *(A. discors, A. cyanoptera, A. crecca)*, Northern Shoveler *(A. clypeata)*, and Northern Pintail *(A. acuta)*.

GADWALL   *Anas strepera*
(Pl. 3)

The male Gadwall, a subtle combination of tan and pale gray, is frequently passed off as "another brown female duck" by casual observers. By looking for a gray duck with a contrasting black

Plate 3. Gadwall: adult male breeding.

rump, birders finally sort through the possibilities and recognize their first male Gadwall. Both male and female Gadwalls have a white speculum, hidden when they are on the water, noticeable as a white patch near the base of the birds' wings in flight.

Gadwalls are medium-sized dabbling ducks. By November, 90 percent of the females are paired up, although they will not leave for northern breeding grounds for another four to five months. If wet conditions provide suitable nesting habitat, Gadwalls are one of the duck species that may remain to breed in southern California.

Recent surveys show Gadwall numbers in North America are steadily increasing.

SIZE: Length 20 in. **ADULT MALE BREEDING:** Sept.–May. Overall gray brown body with black rump and undertail coverts. Pale brown head. White speculum. Black bill. **ADULT FEMALE:** Uniform mottled brown. Smaller head with steeper forehead than female Mallard *(A. platyrhynchos)*. White speculum. Orange bill with black ridge on upper mandible. **STATUS:** Fairly common migrant and winter visitor. Uncommon to fairly common summer resident, especially north of Point Conception. **(B)**

AMERICAN WIGEON   *Anas americana*
(Pl. 4)

Displaying white crowns and foreheads, a group of male ducks paddles around tidal inlets or freshwater marshes on a warm winter day. Formerly known as "baldpates," these American Wigeons live up to their name, as the morning sun reflects like a beacon off their pale foreheads. Listen for the males' call—a whistled "whi-WHEE-who" or "WHEE-who"—sounding like a toy rubber duck being squeezed.

Wigeons graze more than other ducks. They are structurally adapted to eat plant materials and are frequently seen on upland cultivated fields or golf courses nibbling at the grasses.

Wigeons themselves are not adept at diving, but they parasitize other waterbirds that are. On their wintering grounds here in coastal southern California, American Wigeons often associate with coots and diving ducks, from which they steal aquatic plants brought to the surface by these species.

SIZE: Length 20 in. **ADULT MALE BREEDING:** Oct.–June. White forehead and crown, gray head with green crescent sweeping back

Plate 4. American Wigeon: adult male breeding.

from eye. Pale brown breast and sides, white flanks, black undertail coverts. Blue bill tipped with black. **ADULT FEMALE:** Gray head, pale gray brown sides and chest. **STATUS:** Common migrant and winter visitor.

MALLARD   *Anas platyrhynchos*
(Pl. 5)

A duck with a glossy green head, white collar, and chestnut breast is a familiar sight wherever ducks are found. Whether at the edges of a park pond, in a shallow slough, or at the margins of a freshwater marsh, the Mallard is at home almost anywhere. It is particularly accustomed to human haunts, content to join "the breadline" at places where domesticated ducks and geese receive free handouts from visitors.

Widespread throughout the Northern Hemisphere, the Mallard is the ancestor of many breeds of domestic ducks. It is also an abundant wild duck. The task of separating "tame" Mallards from truly wild ones often proves impossible.

The Mallard drake is good sized, one of the largest of the dabblers. (Note how its black, central tail feathers curl upward over the rump.) Like many dabbling ducks, Mallard drakes perform elaborate displays to entice the female. The Mallard drake will

Plate 5. Mallard: adult male breeding.

stretch its head low over the water, then suddenly swim as fast as it can for a short distance, then look up and away from the female.

Mallards vocalize a great deal. The female's descending "QUACK, quack, quack, quack" is evidently an "all's well" announcement to a gathered group of other ducks. Male Mallards give a low muttering "kek-kek-kek" sound.

SIZE: Length 23 in. **ADULT MALE BREEDING:** Oct.–May. Glossy green head, white collar, chestnut breast with dark gray back and pale sides. Blue speculum bordered above and below by white bars. Yellow bill. **ADULT FEMALE:** Uniformly mottled brown, with blue speculum bordered above and below by white bars. Pronounced dark eye line. Orange bill with dark down center of upper mandible. **STATUS:** Common migrant and winter visitor. Fairly common resident, especially in freshwater habitats. **(B)**

BLUE-WINGED TEAL	*Anas discors*
CINNAMON TEAL	*A. cyanoptera*
GREEN-WINGED TEAL	*A. crecca*

(Pls. 6–8)

Along the edges of marshes or rivers, or at the fringes of brackish pools in sloughs, groups of small, brilliantly colored ducks slurp at the duckweed *(Lemna* spp.*)* on the water or nudge their bills into the mud. They put their rumps in the air and submerge their

Plate 6. Blue-winged Teal: adult male breeding.

heads completely to get to aquatic insects at the bottom of the puddles. These teal drakes have exquisite breeding plumages, by which they can be easily recognized.

The male **BLUE-WINGED TEAL** has a distinctive white crescent on the side of its head. This species migrates the longest distance of the three teal, wintering as far south as Brazil from nesting grounds in the prairie pothole regions of southern Canada and the northern United States. In the 1990s, Blue-winged Teal numbers declined drastically because of several dry years on their pothole breeding grounds. With normal rainfall, their population has rebounded, although Blue-winged Teal are never common in our region.

The **CINNAMON TEAL** male stands out because of its allover cinnamon plumage. The Cinnamon Teal breeds only west of the Rocky Mountains, from British Columbia to Mexico City and throughout California, wherever there are suitable wetlands. Although it is commonly seen in southern California, it has the most restricted range of any of the teal species in North America.

The male **GREEN-WINGED TEAL**, the smallest of the three teal, is recognized by a vertical white shoulder stripe. Green-winged Teal have a broad breeding range that extends farther north into Canada and Alaska than that of the other two teal. They winter in the southern United States and are easy to find in our region. Green-winged Teal tend to gather in larger flocks, whereas the

Plate 7 (above). Cinnamon Teal: adult male breeding.

Plate 8 (right). Green-winged Teal: adult male breeding.

other two teal species are more likely to be seen in groups of two or three pairs.

BLUE-WINGED TEAL SIZE: Length 15.5 in. **ADULT MALE BREEDING:** Nov.–June. Gray head with white crescent in front of eye, pinkish brown mottled sides, white flank patch before rear. Dark bill. **ADULT FEMALE:** Mottled brown, white spot at base of bill, dark line through eye, blue speculum. **STATUS:** Uncommon migrant and winter visitor. Rare local resident. **(B*)**

CINNAMON TEAL SIZE: Length 16 in. **ADULT MALE BREEDING:** Oct.–June. Cinnamon head and body. Red eye. Dark, slightly spatulate bill. **ADULT FEMALE:** Mottled brown, plain face pattern, blue speculum. **STATUS:** Common migrant (spring migration Jan.–Apr.). Fairly common winter visitor. Uncommon local resident. **(B)**

GREEN-WINGED TEAL SIZE: Length 14 in. **ADULT MALE BREEDING:** Oct.–June. Head chestnut with green crescent on side of face. Gray body, vertical white shoulder stripe, buffy streak on undertail coverts. Small dark bill. **ADULT FEMALE:** Mottled brown, green speculum, buffy streak on undertail coverts. **STATUS:** Common migrant and winter visitor.

NORTHERN SHOVELER
(Pl. 9)

Anas clypeata

Plate 9. Northern Shoveler: adult male breeding.

Among an assortment of dabbling ducks feeding in shallow water, several have chestnut sides bordered in front and in back by white. Their greenish or purple glossy heads resemble a Mallard's *(A. platyrhynchos),* and they are about the size of a Mallard, but the broad spoonlike bill, wider at the tip than at the middle, distinguishes them from the Mallard. These are Northern Shovelers.

Northern Shovelers are adapted for straining food items from the water, especially tiny crustaceans. Shovelers eat more animal matter than other dabbling ducks, taken by skimming the water with their broad bills. They often swim in small circles of several individuals one behind another, their bills held underwater to scoop up plankton disturbed by the duck in front.

SIZE: Length 19 in. **ADULT MALE BREEDING:** Nov.–May. Head dark green to black, chestnut sides bordered front and back by white, black rump. Large, spatulate black bill. **ADULT FEMALE:** Mottled buffy overall. **STATUS:** Fairly common to common migrant and winter visitor.

NORTHERN PINTAIL
(Pl. 10)

Anas acuta

Plate 10. Northern Pintail: adult male breeding.

A dabbling duck with obvious elongated central tail feathers is a contrast to the other dabblers around it, particularly when its beautiful tail is upended. While it reaches underwater to dislodge seeds and bulbs buried in the mud, the Northern Pintail assumes an unsophisticated pose for such an elegant duck.

Perhaps no other duck is as stunning as the male Northern Pintail in its gray, brown, and white plumage. The long, slender neck combined with the long, pointed tail composes a distinctive silhouette.

Pintails arrive in southern California earlier in fall than most other ducks. By late July, groups begin to assemble along the coast—all still in eclipse plumage. At this stage, the long tail feathers of the drakes are missing.

The Northern Pintail is one of the most abundant waterfowl species in North America, but the overall population is tied to conditions in the prairie pothole region of Canada and the northern United States. When drought affects the prairies, pintail numbers decline. In years of adequate rainfall, they begin to recover.

SIZE: Length 21 in. **ADULT MALE BREEDING:** Nov.–June. Brown head, thin white neck stripe extending onto side of head, white breast, gray body. Long, black central tail feathers. Blue gray bill. **ADULT FEMALE:** Pale brown head, long and slender neck, mottled brown body. **STATUS:** Fairly common migrant and winter visitor.

Diving Ducks (Aythyini)

Diving ducks (tribe Aythyini) are different from dabblers. They need plenty of space to get a running start to take off. Unlike dabbling ducks, they beat their wings rapidly and run along the water to help become airborne. Because they dive for prey—rather than feed near the surface like dabblers—diving ducks are attracted to deep, wide bays and lakes, both for the diving possibilities and the longer stretch of water as a runway.

Diving ducks spend the winter on the deeper bodies of water along southern California's coast. In spring they migrate northward to interior, freshwater lakes to nest.

In southern California, Canvasback *(Aythya valisineria)*, Redhead *(A. americana)*, Ring-necked Duck *(A. collaris)*, and Lesser Scaup *(A. affinis)* are the common species of diving ducks.

CANVASBACK   *Aythya valisineria*
(Pl. 11)

As the ducks float by on a lake near the coast, a birder is lucky to spot among the sleeping shapes one or two that stand out, their backs and sides gleaming pure white. If the ducks have taken their heads out from under their wings for a moment, their long foreheads, which slope smoothly into long, black bills, will confirm that these are Canvasbacks.

Plate 11. Canvasback: adult male breeding.

The Canvasback is a large duck. Its deep chestnut-colored head may lead to confusion with the superficially similar Redhead *(A. americana)*, but the Canvasback's head shape with its sloping forehead is different from that of the rounded forehead of the Redhead.

During the drought in the prairie pothole regions of southern Canada and the northern United States in the late 1980s and early 1990s, Canvasback populations suffered serious declines. They are only just now making a partial comeback.

SIZE: Length 21 in. **ADULT MALE BREEDING:** Oct.–June. Chestnut red head and neck, sloping forehead, black chest, white back and sides. Black bill. **ADULT FEMALE:** Pale brown head, neck, and chest. Sloping forehead, pale gray back and sides. **STATUS:** Uncommon migrant and winter visitor.

REDHEAD
Aythya americana
(Pl. 12)

The same freshwater lakes or brackish lagoons that shelter a few Canvasbacks *(A. valisineria)* may also harbor a flock of Redheads. The Redhead has a rounded, chestnut-colored head. Its back and sides are light gray, not white like that of the Canvasback.

Redheads breed in the western United States and Canada and winter in lagoons along the Gulf of Mexico. In southern Califor-

Plate 12. Redhead: adult male breeding.

nia, small concentrations of breeding Redheads can be found in the Antelope Valley and at the Salton Sea, but rarely along the coast. Occasionally in wet years, Redheads have stayed to nest along the coast at Buena Vista Lagoon, San Joaquin Marsh, and Bolsa Chica.

SIZE: Length 19 in. **ADULT MALE BREEDING:** Oct.–June. Rounded, chestnut-colored head. Black chest, gray back, dark rump. Blue bill with black tip. **ADULT FEMALE:** Tawny brown head and body. **STATUS:** Uncommon migrant and winter visitor. Rare local resident. **(B*)**

RING-NECKED DUCK   *Aythya collaris*
(Pl. 13)

By searching through the flotillas of scaup on deep freshwater lakes and lagoons, one may find a few Ring-necked Ducks.

The plumage pattern of the Ring-necked Duck — dark head and breast and pale sides — is similar to that of the scaup. Up close, though, they are different in several ways. The Ring-necked Duck has a pale white circle before the tip of the bill, a black back, and a whitish vertical mark near the shoulder. Perhaps the Ring-necked Duck should be designated the "Ring-billed" Duck. The ring around the neck, for which it is named, is brown and almost invisible in the field, while the ring around the bill is easy to see.

Plate 13. Ring-necked Duck: adult male breeding.

The Ring-necked Duck is one of the few ducks in addition to the Mallard *(Anas platyrhynchos)* that has increased its wintering population in southern California since the 1930s. Perhaps Ring-necked Ducks do not mind the artificial lakes that have sprung up in urban developments in recent decades, for they like to take shelter under the woody vegetation along the lake margins.

SIZE: Length 17 in. **ADULT MALE BREEDING:** Oct.–June. Dark purple head with slight peak toward rear crown, black chest, pale gray sides with white vertical shoulder mark. White outline at base of gray bill and around end of bill before black tip. **ADULT FEMALE:** Dark gray brown body. Pale eye ring with pale line curving back from eye. White ring around bill before black tip. **STATUS:** Fairly common migrant and winter visitor.

LESSER SCAUP   *Aythya affinis*
(Pl. 14)

On almost any body of moderately deep water near the coast, rafts of ducks with dark heads, pale gray backs, and white sides are accompanied by dark brown ducks with a white patch circling the base of their bills. These are male and female Lesser Scaup, the most common diving duck in our region.

The male Lesser Scaup is elegant in its breeding outfit: dark glossy head, black breast, and white sides. The female looks completely different in a subdued, brownish uniform.

Lesser Scaup winter farther south than most duck species, ranging to Mexico and Central America. They are one of the last to arrive in fall, and they move northward to their nesting grounds in central Alaska and Manitoba later in spring than other ducks.

Plate 14. Lesser Scaup: adult male breeding.

SIZE: Length 16.5 in. **ADULT MALE BREEDING:** Nov.–June. Black head with purplish sheen, black breast, white sides, pale gray back, and black rump. Pale blue bill. **ADULT FEMALE:** Dark brown head with white around base of bill. Dark brown back, grayish sides. Gray bill. **STATUS:** Common migrant and winter visitor.

Sea Ducks and Mergansers (Mergini)

Sea ducks and mergansers are diving ducks that frequent saltwater or brackish water much of the year. Although many nest in interior freshwater wetlands, members of this tribe have a higher tolerance for saltwater than other duck species.

Sea ducks and mergansers feed on fish, mollusks, and other marine organisms, for which they can dive to depths of 40 feet. When they dive, they use their wings for paddling and steering underwater in conjunction with their webbed feet.

In southern California, common wintering sea ducks are the Surf Scoter *(Melanitta perspicillata)*, Bufflehead *(Bucephala albeola)*, and Red-breasted Merganser *(Mergus serrator)*.

SURF SCOTER   *Melanitta perspicillata*
(Pl. 15)

Floating up and over the waves, a group of black ducks rides just beyond the surf break. Oblivious to the spray and the heaving seas, one or two ducks dive, then surface beyond the next threatening wall of water. The rougher the surf, the more they plunge under the waves.

Surf Scoters are equally at home floating in calmer waters, where they feed in sheltered harbors close to piers and jetties. This bird wrests whole mussels, clams, snails, and limpets from pilings and rocks with its powerful bill. Afterward, the muscles of the gizzard go to work digesting the meal, shell and all, with no problem.

In spring, Surf Scoters migrate along the southern California coast by the thousands as they return to northern nesting grounds. They gather to gorge on herring roe at staging areas such as Prince William Sound on the west coast of Alaska. From there, the birds disperse eastward throughout the boreal forests and tundra of Alaska and Canada to nest. In a project that used satellite telemetry to track female Surf Scoters on their West Coast mi-

Plate 15. Surf Scoter: adult male breeding.

gration, researchers found that one individual traveled to breeding grounds as far east as Hudson Bay in Manitoba.

Two other species of scoter occur in our region, particularly north of Point Conception. Although rare, both the White-winged Scoter *(M. fusca)* and Black Scoter *(M. nigra)* may occasionally be found in flocks of Surf Scoters. (See also page 40.)

SIZE: Length 20 in. **ADULT MALE:** All black with white forehead and nape patches. Large orange bill with white spots. **ADULT FEMALE:** Dark brown with two white spots on side of face: one behind bill is slightly vertical, one behind eye is slightly horizontal. Dark bill. **STATUS:** Common migrant and winter visitor. Uncommon summer visitor.

BUFFLEHEAD   *Bucephala albeola*
(Pl. 16)

Plate 16. Bufflehead: adult male breeding.

A small, black-and-white duck swimming in a deep bay, harbor, or brackish lake is likely to be a Bufflehead, the smallest of the sea ducks. The male's large, puffy, mostly white head is distinctive.

Buffleheads dive frequently to feed, then pop up again a short distance away. They look buoyant, bobbing around on the water's surface like toys, then diving again. Male Buffleheads have all-white bodies with black on their heads and backs. Female Buffleheads are completely different: all brown with a small elliptical white patch on the cheek.

Most Buffleheads we see here have nested in central British Columbia and Alberta, where they use abandoned Northern Flicker *(Colaptes auratus)* (a large woodpecker) holes as nesting cavities. Buffleheads' small size may have evolved from their nesting habits, because the Northern Flicker holes are too small for other cavity-nesting ducks.

The Bufflehead is one of the few duck species whose populations have increased since the mid-1950s.

SIZE: Length 13.5 in. **ADULT MALE BREEDING:** Oct–May. Black head with white triangular wedge. White body, black back. Small, gray bill. **ADULT FEMALE:** All brown with white elliptical patch on side of face. **STATUS:** Common migrant and winter visitor.

RED-BREASTED MERGANSER   *Mergus serrator*
(Pl. 17)

A slender waterbird with a long red bill and a ragged reddish brown crest glides along, sometimes with its head poked under the water like a loon surveying the fishing opportunities beneath. The bird doesn't appear to be a duck, nor is it hefty enough to be a loon or cormorant: it is a Red-breasted Merganser.

Mergansers chase down minnows and small fish underwater, catching them in their serrated bills. They have little "sawteeth" pointing backward along the outside edges of the bill, which help the birds hold on to slippery fish.

In our area, Red-breasted Mergansers are typically seen in nonbreeding plumage. In this plumage, the reddish brown of the head merges gradually into the gray of the upper chest. Of the two mergansers that winter in southern California, the Red-breasted Merganser is the one that frequents saltwater, unlike the Common Merganser *(M. merganser)*, which is usually found inland on freshwater.

In spring, small groups of Red-breasted Mergansers migrate north offshore, headed for their breeding grounds in wetlands of the tundra and boreal forests of the high Arctic.

SIZE: Length 23 in. **ADULT MALE BREEDING:** Nov.–May. Glossy green head with ragged crest, white necklace, dark pinkish–streaked breast, gray flanks. Red bill. **ADULT MALE NONBREEDING (JUNE–OCT.)**

Plate 17. Red-breasted Merganser: adult female.

AND FEMALE: Crested head reddish brown, merging gradually with grayish chest. **STATUS:** Common migrant and winter visitor. Uncommon summer visitor.

Stiff-tailed Ducks (Oxyurini)

The only common member of this tribe in the United States is the widespread Ruddy Duck *(Oxyura jamaicensis).*

RUDDY DUCK   *Oxyura jamaicensis*
(Pl. 18)

A small brownish duck with a white cheek and a tail held stiffly at an upward angle swims in good-sized flocks at almost any watery venue, except the open ocean, in southern California. On a lake, slough, estuary, or harbor, the sight of little brown ducks with stiff tails triggers an immediate response from a birder—it's a Ruddy Duck. (Caution: Some Ruddy Ducks float without raising their tails.)

When threatened, Ruddy Ducks would rather dive than fly.

Plate 18. Ruddy Duck: adult male breeding.

They behave surprisingly like grebes: disappearing slowly under the water's surface when necessary.

The Ruddy Duck is one of the few duck species that does not molt into eclipse plumage in summer. Instead, the drake wears a drab, nonbreeding plumage during the winter season (when most other ducks are already in nuptial garb), then dons a rich, chestnut red coat and an electric blue bill come spring.

Although Ruddy Ducks are mostly winter visitors, some stay to breed in our area if they can find suitable, tule-bordered marshes and ponds.

SIZE: Length 15 in. **ADULT MALE BREEDING:** Mar.–Aug. Chunky duck with black cap, pure white cheek, reddish brown back, sides, and breast. Bright blue bill. **ADULT MALE NONBREEDING:** Sept.–Mar. Drab brown with white cheek. **ADULT FEMALE:** Drab brown with dark line across white cheek. **STATUS:** Common migrant and winter visitor. Fairly common resident. **(B)**

LOONS
(Gaviidae)

Loons are large, elongated birds adapted to aquatic habitats. They are the ultimate diving machine: webbed toes; legs set far back on the body; long, streamlined shape; myoglobin in the muscles to store oxygen for long dives; dense bones to help the bird submerge. On land, they shuffle along helplessly on their bellies.

Loons ride low on the water's surface. They have short necks and sharp, thick bills to grasp fish. They frequently put their heads partially underwater to peer beneath the surface for prey. Sighting a fish, they lunge forward in a dive. Loons can chase small fish to a depth of over 250 feet, staying under for up to a minute.

Loons, although beautiful in breeding plumage, are in dull, nonbreeding garb for much of their stay in our region.

Three species of loon are found along the southern California coast: the Red-throated Loon *(Gavia stellata)*, Pacific Loon *(G. pacifica)*, and Common Loon *(G. immer)*.

RED-THROATED LOON   *Gavia stellata*
(Pl. 19)

Plate 19. Red-throated Loon: adult nonbreeding.

A long, slender bird with a short neck and a delicate, slightly up-tilted bill swims low on the water in sheltered harbors or floats in shallow surf just off the beach. The overall impression is that of a pale gray bird with a plain white face.

This is the Red-throated Loon, the smallest member of the loon family. As a result of the slight upward curve of its lower mandible, the Red-throated Loon appears to hold its bill at an upturned angle — not to be confused with the cormorant, which has a much longer neck, thicker bill, and holds its bill at an upward slant also. In contrast, both Common and Pacific Loons (*G. immer* and *G. pacifica*) hold their bills straight.

Red-throated Loons spend the winter along the Pacific coast from southern Alaska to Baja California. Although not as abundant as Pacific Loons, they migrate north in fair numbers past our shores in March and early April on their way to breed on the tundra and Arctic seacoasts of North America.

SIZE: Length 25 in. **ADULT BREEDING:** Apr.–Nov. Sexes similar. Back dark or brownish gray. Face and neck pale gray, chestnut patch on throat. Black bill. **ADULT NONBREEDING:** Oct.–Apr. Sexes similar. Upperparts light gray with back variably speckled with white dots. Only slight contrast between upperparts and underparts. Pale bill. **STATUS:** Fairly common to common migrant and winter visitor. Rare summer visitor.

PACIFIC LOON   *Gavia pacifica*
(Pl. 20)

On April days, if a strong onshore breeze blows the birds close to shore, a birder standing at Point Vicente, Point Dume, or Goleta Point, will see flock after flock of Pacific Loons flying past. The peak passage begins at sunrise when, at certain locations, thousands of loons pour by the coast in several hours.

The Pacific Loon is the most numerous of the North American loons and the most abundant loon in spring migration in southern California's nearshore waters. If the loons pass close enough, you may glimpse their stunning breeding plumage of white back patches, silver gray napes, and black throats as the birds beat their way north with powerful wings.

In winter, however, Pacific Loons are not a common sight. They prefer deep water and do not usually come close to shore.

Plate 20. Pacific Loon: adult breeding.

A Pacific Loon floating off the end of a pier or jetty—from which they are sometimes seen in winter—will be in drab nonbreeding plumage. (See also pages 38–39.)

SIZE: Length 25 in. **ADULT BREEDING:** Apr.–Oct. Sexes similar. Crown and paler nape soft gray. Rows of white patches on black back. Black throat set off with white streaks. Black bill. **ADULT NONBREEDING:** Sept.–Mar. Sexes similar. Dark cap comes down to eye. Dark of nape and hindneck contrasts cleanly with white face and foreneck, making a sharp border. Sometimes has dark chin strap. Gray bill. **STATUS:** Common spring migrant. Uncommon fall migrant and winter visitor offshore. Rare summer visitor.

COMMON LOON   *Gavia immer*
(Pl. 21)

A large, dark gray seabird with a heavy, daggerlike bill floats in our harbors, bays, and inlets, seeming to blend with the color of the gray sea on a cloudy day. Occasionally, the bird puts its head underwater to watch for fish as it glides along.

In southern California, the Common Loon is seen in drab winter plumage, but on its breeding grounds, this gorgeous bird has a black head, white necklace, and checkered black-and-white

Plate 21. Common Loon: adult nonbreeding.

back. The subject of myth and legend, the Common Loon is the classic loon that nests on forested lakes in Alaska and northern Canada, as well as the northern tier of states in the continental United States. As such, it is the most southerly nester of the three loons.

The yodeling call of the Common Loon—the famous "song" of the loon—is seldom heard in southern California; the birds are silent on their wintering grounds.

Although Common Loons are frequently seen in our harbors during the winter, in spring migration when Red-throated and Pacific Loons *(G. stellata* and *G. pacifica)* are passing in numbers offshore, relatively few Common Loons accompany them. Most Common Loons take a more inland route, stopping at large reservoirs and freshwater lakes on their way north.

SIZE: Length 32 in. **ADULT BREEDING:** Mar.–Oct. Sexes similar. Black head, white necklace, back checkered with small spots. Black bill. **ADULT NONBREEDING:** Sept.–Mar. Sexes similar. Dark gray above, pale white throat and neck, white around eye. Traces of dark collar on sides of neck. Pale bill. **STATUS:** Common winter visitor and fairly common to uncommon migrant. Rare summer visitor.

GREBES
(Podicipedidae)

Grebes spend most of their lives right on the water: feeding, sleeping, and courting. They nest on floating platforms of vegetation.

Grebes resemble loons in body structure and habits, but they differ in several respects. Grebes possess lobed toes, not webbed ones, and they occupy a wider variety of aquatic habitats than loons. Like many waterbird species, grebes perform an amazing feat: they can adapt to both salt- and freshwater. Most grebes breed on inland freshwater lakes and marshes, then migrate to the sea coast to spend the winter. They do not migrate the extreme distances that loons do, however, and they do so at night, not during the day like loons.

Grebes have the unique habit of eating their own body feathers. Researchers speculate this may protect the stomach from indigestible parts of the grebe's prey. It also aids in the formation of pellets, which are regurgitated from time to time and contain the undigested bones of fish.

Three small grebes, Pied-billed *(Podilymbus podiceps)*, Horned *(Podiceps auritus)*, and Eared *(P. nigricollis)*, and two large grebes, Western *(Aechmophorus occidentalis)* and Clark's *(A. clarkii)*, are found along the southern California coast.

PIED-BILLED GREBE   *Podilymbus podiceps*
(Pl. 22)

A brownish, compact bird with a short neck and a white rump lurks at the margins of sheltered harbors, ponds, or sloughs near the coast but avoids the open ocean. The Pied-billed Grebe also glides between the reeds in freshwater marshes, looking from a distance like a little brown duck. Its unique, blunt bill with a downcurved upper mandible distinguishes the Pied-billed from other grebes. The bill is marked in breeding plumage by a dark band around it, hence the name "pied." Listen for the loud descending "caow, caow, caow" call coming from a Pied-billed Grebe hiding in the tules.

The Pied-billed Grebe is adept at sinking slowly out of sight beneath the water's surface when disturbed, expelling air from

Plate 22. Pied-billed Grebe: adult nonbreeding.

internal air sacs, which allows it to sink underwater. A grebe observed doing this submarine behavior is usually a Pied-billed.

Male Pied-billed Grebes are surprisingly aggressive year-round toward their own species as well as others. There is evidence that when Pied-billeds threaten to attack, other birds leave the immediate area.

SIZE: Length 13 in. **ADULT BREEDING:** Feb.–Sept. Sexes similar. Similar to adult nonbreeding, but with black vertical band across white bill. **ADULT NONBREEDING:** Sept.–Mar. Sexes similar. Overall brownish. White rump. Flesh-colored bill. **STATUS:** Common migrant and winter visitor. Fairly common resident. **(B)**

HORNED GREBE   *Podiceps auritus*
(Pl. 23)

Among the wintering waterbirds that swim in the bays near San Diego or Newport, or dive in harbors off Ventura and Santa Barbara, a small gray-and-white bird with a thin neck and a pointed bill can be spotted infrequently.

The Horned Grebe, the least common of our grebes, can be distinguished from the similar Eared Grebe *(P. nigricollis)* in nonbreeding plumage by its clean white cheek and foreneck. The Eared Grebe has a dirty gray cheek and foreneck.

Plate 23. Horned Grebe: adult nonbreeding.

Horned Grebes nest farther to the north in Alaska and northwestern Canada than the other grebe species, which winter along our coast.

SIZE: Length 14 in. **ADULT BREEDING:** Apr.–Aug. Sexes similar. Solid yellow crescent above black cheek, bright rufous neck. Dark bill. **ADULT NONBREEDING:** Sept.–Mar. Sexes similar. Black cap ends in horizontal line separating it from white cheek. White throat and breast. Gray bill. **STATUS:** Uncommon to fairly common migrant and winter visitor.

EARED GREBE   *Podiceps nigricollis*
(Pl. 24)

A small gray bird with a smudgy gray face and neck and a white, fluffy rump is a frequent sight anywhere along the coast in winter. The Eared Grebe, more versatile than its close kin the Horned Grebe *(P. auritus),* is satisfied with a variety of habitats. It associates with coots and ducks in freshwater marshes or brackish sloughs but also ventures to nearshore waters among loons and cormorants.

Plate 24. Eared Grebe: adult nonbreeding.

Eared Grebes have tiny, pointed, slightly upturned bills, and their heads come to a peak on top; Horned Grebes have longer, straighter bills and flatter heads.

The migration ecology of Eared Grebes is interesting. After breeding inland, they gather on saline lakes such as Mono Lake and Great Salt Lake to molt and feed for several months, thus accumulating large fat deposits before flying to the coast to winter. They feed on plentiful brine shrimp, sometimes postponing their migrations until too late in the season. On the night of December 10, 1991, a snowstorm in southern Utah forced down thousands of migrating grebes en route to wintering spots in southern California and Mexico. Becoming disoriented by lights from towns and intersections, the grebes landed on highways or crashed to the ground, where many were captured alive and released on bodies of water nearby. This points to the role of weather in Eared Grebe migrations: the birds must weigh the dangers of an early snowstorm against the benefits of exploiting abundant food sources at their staging grounds (gathering areas where birds feed or molt, usually in preparation for migration).

SIZE: Length 13 in. **ADULT BREEDING:** Apr.–Sept. Sexes similar. Overall black with reddish highlights. Yellow plumes on sides of head form "ears." **ADULT NONBREEDING** Oct.–Mar. Sexes similar. Dark cap ends in indistinct line separating it from grayish white cheek, gray neck. Whitish rump. Gray bill. **STATUS:** Common migrant and winter visitor. Rare summer visitor and resident. **(B*)**

WESTERN GREBE   *Aechmophorus occidentalis*
CLARK'S GREBE   *A. clarkii*
(Pls. 25, 26)

The graceful, swanlike appearance of Western and Clark's Grebes makes them the stars of the grebe family. Until fairly recently, Clark's Grebes were considered a color phase of the Western Grebe. In 1985, ornithological research determined that although their ranges overlap and they have similar habits and structure, Western and Clark's Grebes rarely interbreed.

Both Western and Clark's Grebes possess a mechanism that allows the neck to push the head forward, as though it were a spear. This ability for rapid neck extension is shared by some of the herons, but no other species of grebe. It works well for chasing fish underwater.

Plate 25. Western Grebe: adult nonbreeding.

On a winter day when the sea is quiet, rafts of hundreds of Western Grebes—a few Clark's Grebes among them—float calmly out beyond the surf line. Usually they are dozing, with their necks drawn back between their wings and their bills tucked to one side. When one of them awakes, it may roll over on its side to preen, with one ungainly long leg waving in the air.

Before Western and Clark's Grebes depart for inland nesting sites in the spring, they perform spectacular courtship displays. The most well known is a sort of water dance, in which a pair of grebes arch their necks, rear up on their feet, and patter rapidly across the water's surface side by side. After about 100 yards, they suddenly stop and dive underwater.

Plate 26. Clark's Grebe: adult breeding (Brad Sillasen).

To the untutored eye, Western and Clark's Grebes look exactly alike. Both have long, slender necks with pure white on the front and dark on the back. Both have sharp, pointed bills. The differences between the two are subtle. The Clark's Grebe has more white on the face, a brighter orange bill, and a different call. The easiest way to tell them apart is bill color: greenish yellow in the Western Grebe and bright orange yellow in the Clark's.

Another difference between the two grebes is in their calls. The Western gives a two-syllable "cree-creet?", whereas the Clark's is a one syllable "creet?" Grebes are vocal, so learning their calls is a help in identification.

WESTERN GREBE SIZE: Length 25 in. **ADULT BREEDING:** Feb.–Sept. Sexes similar. Similar to adult nonbreeding, but plumage more vivid. **ADULT NONBREEDING:** Sept.–Feb. Sexes similar. Large, slender waterbird with long neck, black hindneck, and white foreneck. Black of cap surrounds eye. Greenish yellow bill. **STATUS:** Common migrant and winter visitor. Uncommon summer visitor.

CLARK'S GREBE SIZE: Length 25 in. **ADULT BREEDING:** Feb.–Sept. Sexes similar. Similar to adult nonbreeding, but white of face surrounds eye. **ADULT NONBREEDING:** Sept.–Feb. Sexes similar. Similar to Western Grebe, except bright orange yellow bill. **STATUS:** Fairly common migrant and winter visitor. Uncommon summer visitor.

SHEARWATERS (Procellariidae)

Like their relatives the petrels, fulmars, and prions, the shearwaters skillfully ride wind currents as they roam the oceans of the world. All are pelagic, spending their entire lives at sea and only briefly visiting land during the breeding season.

Shearwaters are members of a group of seabirds known as "tubenoses," in reference to their external nostrils, which form twin horny tubes at the base of the upper bill, enhancing the birds' ability to smell. This well-developed sense of smell guides them toward prey and helps them locate their young on remote nesting islands at night.

How do pelagic birds such as the shearwaters, which live at sea, survive without drinking fresh water? The tubenoses, like the cormorants, pelicans, gulls, terns, and other seafaring birds, possess a pair of salt glands located near the eye sockets and above the bill. The excess salt from ingested seawater is extracted from the bird's blood by the salt glands, where it is channeled to the nasal cavities. The concentrated salt solution is then excreted through the nostrils.

Shearwaters get their name from their habit of "shearing the waves" in graceful glides and dips, flying with long pointed wings held stiffly just inches above the ocean swells. They have a distinctive flight pattern of several quick flaps alternating with a prolonged glide.

Two species of shearwater are common off southern California: Sooty Shearwater *(Puffinus griseus)* and Black-vented Shearwater *(P. opisthomelas).*

SOOTY SHEARWATER   *Puffinus griseus*
(Pl. 27)

Plate 27. Sooty Shearwater: adult.

In August or September, a birder driving north along the coast should plan to stop at the Santa Maria River estuary or at the cliffs at Shell Beach. From these vantage points, and with the help of a good spotting scope, a mass of birds flying in an endless swirl over the ocean can often be descried. As the birds at the front of the flock alight to feed for a moment, those at the rear overtake them and pass on above. The entire mass, sometimes tens of thousands of individuals, continues to move forward in this leapfrog fashion. The birds are following schools of anchovies, a favorite food of the Sooty Shearwater.

From a boat in the Santa Barbara Channel, you may get a closer view. As the boat approaches, the Sooty Shearwater will take off in a running pitter-patter across the surface. (At first glance, it resembles one of the brownish, immature Heermann's Gulls *[Larus heermanni]*, which are common offshore in summer.) Then, as the bird gains altitude, it begins to cruise effortlessly above the water, its glide broken now and then by several quick flaps. First one wing and then another grazes the crest of the wave before the Sooty Shearwater disappears into the trough of the swell and is gone.

Many of the Sooty Shearwaters found off California nest in the Southern Hemisphere in New Zealand. They travel an incredible figure eight migration: from New Zealand eastward to the Peru Current in winter, northwestward to the western Pacific in spring, then eastward to the eastern North Pacific during summer, and finally southwestward to New Zealand in fall. Most are nonbreeders and travel nearly 25,165 miles (40,500 kilometers) annually in enormous flocks searching for prey. During El Niño–Southern Oscillation events, when waters off the California coast are warmer and harbor less prey, Sooty Shearwaters remain in the cooler central North Pacific, avoiding California shores. When sea surface temperatures become colder again, they return to feed in waters off the California coast.

SIZE: Length 17 in. **ADULT:** Sexes similar. Dark chocolate brown over all with silvery underwings. Dark bill. **STATUS:** Common visitor spring through fall, particularly north of Point Conception. Uncommon winter visitor.

BLACK-VENTED SHEARWATER *Puffinus opisthomelas*
(Pl. 28)

When a feeding frenzy starts to build in inshore waters, and seabirds come flocking from every direction to plunge and grab and snatch at the seething mass of fish beneath the surface, look for the black above and white underneath pattern of the small

Plate 28. Black-vented Shearwater: adult (Walter Wehtje).

Black-vented Shearwater among the pelicans, cormorants, and gulls. As large fish drive the anchovies to the surface, it is tantamount to placing them on a serving platter for the shearwaters and other hungry seabirds.

If the feeding frenzy is close enough to shore, a birder with a spotting scope can identify the Black-vented Shearwaters. The Black-venteds fly with faster wing beats and shorter glides than Sooty Shearwaters *(P. griseus)*. Wings beating furiously, they approach the mob of diving birds and plop down in the water. Once landed, the Black-venteds bob and stab at the swarming fish.

Ninety-five percent of the world's Black-vented Shearwaters breed solely on Natividad Island off the west coast of Baja California, where they lay their eggs in burrows. After nesting here and on a few adjacent islands, they venture north and south remaining fairly close to shore to forage on Northern Anchovies and Pacific Sardines *(Sardinops sagax)*. Look for Black-venteds from Point Conception south to San Diego with most birds present from fall through winter.

SIZE: Length 14 in. **ADULT:** Sexes similar. Dark brown, appearing black above. White below with dark undertail coverts. Dark bill. **STATUS:** Fairly common to uncommon late fall and winter visitor. May linger through early spring.

PELICANS
(Pelecanidae)

The enormous bill, bulky body, and short legs of the pelicans make them look awkward on land. Despite the pelicans' odd proportions, they are agile when swimming and flying. Pelicans have a totipalmate foot: all four toes are connected with webbing. Also, pelicans have a flexible pouch of bare skin — a gular pouch — which hangs from their lower mandible. The combination of the paddlelike foot and the expandable pouch allows pelicans to be extremely adept at catching fish.

Two species of pelican are found in coastal southern California: the American White Pelican *(Pelecanus erythrorhynchos)* and the Brown Pelican *(P. occidentalis).*

AMERICAN WHITE PELICAN   *Pelecanus erythrorhynchos*
(Pl. 29)

On a winter morning when the tide is full in Morro Bay, a flock of large white birds with black wing tips takes shape out of the fog and slowly sails in for a landing. The birds are so large that even with the naked eye, you can pick out the pelican shape and the way their thick orange yellow bills rest on their chests. When the American White Pelicans come to roost on the nearby sandspit, they settle down and hide their bills in their wings, each one resembling a crumpled pile of white laundry.

American White Pelicans feed in protected bays, harbors, and freshwater lakes, never on the open ocean. They hunt in groups, instead of individually like the Brown Pelican *(P. occidentalis).* Several American White Pelicans will gather in a semicircle and herd a school of fish toward shore. They beat the water with their wings to nudge the fish onward. Once the pelicans' prey has been driven into shallow water, the birds scoop the fish into their large pouches from a sitting position. On other occasions, they form two parallel lines and swim toward one another, herding the fish between them.

American White Pelicans have a 9-foot wingspan, almost as wide as a California Condor *(Gymnogyps californianus).* They are expert fliers, traveling miles to and from nesting colonies on in-

Plate 29. American White Pelican: adult nonbreeding.

land lakes in the Great Basin. It is a strange and beautiful sight to see a mass of White Pelicans wheeling high above mountains and deserts, miles from any body of water.

SIZE: Length 62 in. **ADULT BREEDING:** Feb.– June. Sexes similar. Similar to adult nonbreeding, but a raised knob develops halfway along the upper mandible. **ADULT NONBREEDING:** Sept.–Feb. Sexes similar. Large, white bird with black flight feathers (primaries and secondaries). Orange yellow bill and pouch. **STATUS:** Common fall and winter visitor and rare summer visitor at Morro Bay, uncommon to rare migrant and winter visitor elsewhere.

BROWN PELICAN   *Pelecanus occidentalis*
(Pl. 30)

Groups of Brown Pelicans fly parallel to the coast, cruising in a straight line several feet above the water. Using the updraft created by the cresting wave, they set their wings at the same horizontal level, flapping only occasionally. When they do flap, it is in unison. Their silhouettes—composed of long bills, wide wings, and short tails—are a classic southern California sight from shore.

Brown Pelicans gather on sandspits, beaches, and wharves. In a group, the adult birds stand out because they have white heads, whereas the juveniles are entirely grayish brown.

The Brown Pelican forages in saltwater and can plunge-dive from as high as 60 feet. Extending its neck and holding its wings back, the pelican hits the water head first with a great splash. Air sacs protect its skull from the force of the impact. The bird uses its bill to scoop up a fish with a mouthful of seawater. As the water drains from the bill, the pelican rearranges the fish, which it catches crosswise, in order to swallow the fish lengthwise.

Plate 30. Brown Pelican: adult nonbreeding.

When a Brown Pelican fishes, gulls and terns often harass it, watching for a chance to steal the fresh catch. A Heermann's Gull (*Larus heermanni*) will rob the pelican of its fish by standing on the pelican's back and pecking at its bill, forcing the bird to open its pouch. Or, the Heermann's Gull swims close enough to snatch the pelican's fish just as it is being rearranged for the proper swallow. This is called kleptoparasitism, a long word meaning "stealing of prey" (see fig. 12).

Brown Pelicans nest on Santa Barbara and Anacapa Islands in large colonies (5,000 to 7,000 nests annually). After they leave their nest sites in late June, they join other Brown Pelicans, which have nested off Baja California, to wander north along the Pacific coast.

The Brown Pelican was in serious trouble in the early 1970s as a result of eggshell thinning caused by ingestion of the pesticide

DDT. With the banning of DDT, the Brown Pelican has made a steady comeback; however, California Brown Pelicans (*P. o. californicus*) are still listed as endangered by both the state and federal governments because, though 90 percent of the subspecies nests south of the Mexican border, the U.S. population, those nesting on Santa Barbara and Anacapa Islands, is vulnerable to fluctuations. For example, in years when the warm water associated with El Niño–Southern Oscillation events results in a scarcity of anchovies, California Brown Pelicans are unable to find food for their nestlings, which may then perish. In addition, Brown Pelican colonies on the islands are threatened by human disturbance. A recent concern has been the impact of the bright lights used at night by market squid fishing boats.

SIZE: Length 51 in. **IMMATURE:** Sexes similar. Gray brown upperparts and chest, white belly. **ADULT BREEDING:** Dec.–Aug. Sexes similar. Yellow crown, dark chestnut hindneck, dark gray back and underparts. Pouch bright red at base. **ADULT NONBREEDING:** Aug.–Jan. Sexes similar. White head and neck, dark gray back and underparts. Pink bill. **STATUS:** Common year-round visitor on mainland coast, largest numbers present in summer and early fall. Resident on Anacapa and Santa Barbara Islands. **(B*, FE, SE)**

CORMORANTS
(Phalacrocoracidae)

Cormorants are large, glossy black seabirds with long necks. Like loons and grebes, they are superb divers and swimmers. Like pelicans, they have a gular pouch, and all four toes are connected with webbing.

Cormorants sit erect on buoys, cliffs, nearshore rocks, or abandoned platforms. Their stiff tails help them cling to the steepest surfaces. When they leave seaside perches, they fly with rapid wing beats and neck extended, eventually skidding to a halt when they hit the water. To dive, cormorants leap forward from a sitting position and swim underwater with swiftly paddling feet.

Between dives to hunt for fish, cormorants float low in the

water, often looking partially submerged. The bill is held at an upward slant, a characteristic pose.

After a session of underwater foraging, cormorants climb onto a convenient rock or wharf piling and sun themselves with both wings spread out. This spread wing posture helps the birds dry and rearranges the intricate structure of their wing feathers in order to keep them waterproof.

In spring breeding plumage, some species of cormorants grow small white plumes; the bare skin around their pouches turns orange, turquoise, red, or buff.

Three species of cormorant frequent southern California: Brandt's *(Phalacrocorax penicillatus)*, Double-crested *(P. auritus)*, and Pelagic *(P. pelagicus)*.

BRANDT'S CORMORANT   *Phalacrocorax penicillatus*
(Pl. 31)

Looking out to sea from a good vantage point on shore, a birder notices a large aggregation of black seabirds with snakelike necks sitting on the water. The birds dive, then pop up again, splashing actively. If they are following a school of fish, the birds to the rear continually fly up to the front of the group and land. These are Brandt's Cormorants, the most common cormorant in the Southern California Bight.

Plate 31. Brandt's Cormorant: adult nonbreeding.

Brandt's Cormorant is endemic (native) to the Pacific coast. It was named after a Russian naturalist, J. F. Brandt, who first described the bird in 1837 from a specimen he studied at the Zoological Museum of St. Petersburg, collected by one of the Russian expeditions along the Pacific coast in the 1830s.

Brandt's Cormorants do not migrate, but they move around depending upon food availability. During the breeding season, they rely upon the California Current and the prey attracted by the upwelling of colder waters in spring. When El Niño–Southern Oscillation events cause warmer water temperatures, Brandt's Cormorant populations decline.

From afar, it is difficult to tell Brandt's and Double-crested Cormorants *(P. auritus)* apart. They are both large, black cormorants of about the same size. At closer range, the Brandt's buffy band at the base of its throat pouch can be distinguished from the Double-crested's orange throat pouch.

SIZE: Length 34 in. **ADULT BREEDING:** Apr.–June. Sexes similar. Similar to adult nonbreeding, but with turquoise blue at base of bill, white plumes scattered along the sides of the head and neck. **ADULT NONBREEDING:** June–Mar. Sexes similar. Glossy black all over with buffy band at base of throat. Dark bill. **STATUS:** Common migrant and winter visitor. Common resident on the mainland north of Point Conception and on all the Channel Islands except Santa Catalina. Rare local resident in San Diego County and on the South Coast of Santa Barbara County. Uncommon away from breeding areas in summer. **(B)**

DOUBLE-CRESTED CORMORANT   *Phalacrocorax auritus*
(Pl. 32)

Sunning itself with wings partially extended at the end of a pier, or swimming in the harbor of a coastal town, the Double-crested Cormorant is accustomed to humans; it's the most frequently seen cormorant anywhere along the mainland.

The Double-crested Cormorant is named for its twin white head tufts, like bushy white eyebrows, acquired during the breeding season. A more reliable field mark all year long is the bright orange gular pouch or throat.

For years the species has been the subject of controversy because fishermen and aquaculture operations perceive the birds as

Plate 32. Double-crested Cormorant: immature.

unfair fishing competition; however, although Double-crested Cormorants prey exclusively on fish, they are opportunistic and will take many varieties including those undesirable to humans.

Double-crested Cormorants are the only North American cormorant to occur in large numbers in the interior as well as on the coast. Any cormorant roosting in eucalyptus trees near a lagoon or preening on a dock beside an inland freshwater lake is a Double-crested. Double-crested Cormorant populations are increasing in North America, to the point where their conspicuous colonies have become somewhat of a nuisance in certain parts of the country.

SIZE: Length 33 in. **IMMATURE:** Sexes similar. Brown with whitish breast. Orange gular pouch and bill. **ADULT BREEDING:** Mar.–May. Sexes similar. Similar to adult nonbreeding, but with two white tufts on either side of head. **ADULT NONBREEDING:** June–Feb. Sexes similar. Allover glossy black, orange gular pouch, black bill. **STATUS:** Common migrant and winter visitor. Fairly common local resident on the mainland and many of the Channel Islands. **(B)**

PELAGIC CORMORANT   *Phalacrocorax pelagicus*
(Pl. 33)

To see Pelagic Cormorants at their nesting sites, visit Santa Cruz or Santa Rosa Island, or scan the cliffs on the mainland near Shell Beach or San Simeon.

Clinging to the sides of the most precipitous sea cliffs and islets, Pelagic Cormorants eschew more level ground, where the Double-crested and Brandt's Cormorants *(Phalacrocorax auritus and P. penicillatus)* have staked out rookeries. Instead, pairs choose the most inaccessible niches down the cliff sides for nesting in loose colonies. From there, they dive into the currents at the base of the cliffs, searching for fish that lurk among the rocks.

Pelagic Cormorants are noticeably smaller and slimmer than the other cormorants. Their heads are the same width as their necks, giving them a thin, "pencil" look when they stretch out in flight. They are easiest to identify in spring, when the adult birds sport a pair of white flank patches.

Plate 33. Pelagic Cormorant: adult non-breeding.

The Pelagic Cormorant is a western species, resident along the Pacific coast from Alaska south to Point Conception. It is the least common of the cormorants in our region, being at the southern extent of its breeding range here.

SIZE: Length 28 in. **ADULT BREEDING:** Feb.–May. Sexes similar. Similar to adult nonbreeding, but white plumes scattered on neck, two white flank patches, red at base of bill. **ADULT NONBREEDING:** June–Mar. Sexes similar. Glossy black all over including face. Gray bill. **STATUS:** Fairly common resident north of Point Conception and on the Channel Islands south to Santa Barbara Island. Uncommon to fairly common migrant and winter visitor elsewhere. Rare summer visitor south of Point Conception. **(B)**

HERONS, EGRETS, and BITTERNS (Ardeidae)

Herons and egrets are long-necked, long-legged wading birds. The taller species, such as the Great Blue Heron *(Ardea herodias)* and the Great Egret *(A. alba),* stand motionless along the shores of ponds, lakes, streams, and estuaries hunting for prey.

Members of the heron family have four slender, spreading toes that support them on muddy ground. On each foot the middle front toe has a serrated edge used as a comb. Herons grow special powder-down feathers in three areas of their body. These feathers are brittle, disintegrating into a talcumlike dust when touched by the bird's bill as it preens. By using the clawed toe, the bird distributes the dust as a cleaning agent throughout its plumage.

When herons fly, they hold their head back on their shoulders with the neck in an S curve. (Great Blue Herons are sometimes mistaken for cranes, which fly with neck extended.) As the heron flies, it takes deep, slow wing beats, and its long legs stick out behind the tail.

Egrets are named for the beautiful body plumes that drape from their head, neck, and back in breeding plumage: *aigrettes* (a French word meaning "plumes"). The white plumes, prized by women in the late 1800s and early 1900s to adorn their hair, hats, and dresses, nearly caused the demise of the Snowy Egret *(Egretta thula)*, and, to a lesser extent, the Great Egret. Plume hunters ravaged the egret colonies, killing thousands of birds a year to obtain fresh aigrettes for the plume trade. It was the indignant response of preservationists, who urged laws to stop the slaughter, that spawned the early campaigns of the National Audubon Society and the American Ornithologists' Union. Eventually, interest in feathery hats subsided. The preservationist effort was one of the first conservation statements on behalf of a threatened bird species in U.S. history.

Herons feed on a variety of prey: frogs, mice, fish, rats, lizards, shrimp, crabs, ground squirrels, and pocket gophers. Sometimes they stab prey with their sharp bill first, but usually they grasp it and swallow it all at once.

Five species of heron can be found in southern California: Great Blue Heron, Great Egret, Snowy Egret, Green Heron *(Butorides virescens)*, and Black-crowned Night-Heron *(Nycticorax nycticorax)*. Also in the heron family, but not treated in this book, are the Cattle Egret *(Bubulcus ibis)*, American Bittern *(Botaurus lentiginosus)*, and Least Bittern *(Ixobrychus exilis)*. The Cattle Egret is a rare visitor to the coast in fall and winter. Bitterns are found in marshes, where they hide in dense vegetation and are difficult to see.

GREAT BLUE HERON   *Ardea herodias*

(Pl. 34)

In the topmost branches of groves of eucalyptus or cypress trees at Point Loma, Channel Islands Harbor, Goleta Beach, or Morro Bay State Park, the 4-foot-tall silhouette of a Great Blue Heron can be seen against the sky. Upon closer inspection, a dozen or more of these incongruous forms may take shape, camouflaged among the branches. Some are perched near their platform nests. Others fly overhead, their 6-foot wingspan and dangling legs making them unlikely candidates for nesting in such high trees. The statuesque gray-and-blue figure of a solitary Great Blue Heron is a common sight at a coastal estuary, at the end of a wharf, or even on an unoccupied boat at a harbor. When disturbed, the heron will slowly flap off, giving a loud, harsh croak in protest.

In addition to fish, the Great Blue Heron may hunt gophers

Plate 34. Great Blue Heron: adult.

and mice in fields and pastures. Watching one of these birds successfully devour oversized prey is enlightening. Due to its expandable gullet, the Great Blue Heron has no difficulty eating gophers or large fish. But it takes time before the prey is completely swallowed, so the heron's neck swells in an unsightly lump for several minutes (see also page 47).

SIZE: Length 46 in. **ADULT:** Sexes similar. A tall wading bird with whitish face and short, black head plumes. Pale gray neck and blue gray body. Pointed, yellowish bill. **STATUS:** Common migrant and winter visitor. Fairly common local resident. **(B)**

GREAT EGRET
Ardea alba

(Pl. 35)

When the fog begins to lift from the salt marsh at San Elijo or Carpinteria, a long-legged, all-white heron emerges like a ghost from the mist. Motionless, this tall wader stands with neck stretched forward, its thick, sharp bill held ready to strike. When the unlucky frog or fish swims close enough, the Great Egret

Plate 35.
Great Egret:
adult.

makes a sudden lunge, evidence that its hunt for breakfast is serious business.

After the Great Blue Heron *(A. herodias)*, the Great Egret is the largest heron in our area. Its tall stature, long yellow bill, and black legs and feet distinguish it from the smaller Snowy Egret *(Egretta thula)*.

SIZE: Length 39 in. **ADULT:** Sexes similar. Tall, long-necked, all-white heron. Black legs and feet. Yellow bill. **STATUS:** Fairly common migrant and winter visitor. Rare local resident, occasionally nesting in Great Blue Heron colonies. **(B*)**

SNOWY EGRET   *Egretta thula*
(Pl. 36)

A group of white wading birds at a coastal lagoon is an arresting sight. The graceful breeding plumes on their necks and backs wave gently in the breeze. From time to time, two or three of the group make little jumps into the air, flicking their wings to scare prey to the surface of the water. Another walks slowly along the

Plate 36. Snowy Egret: adult breeding.

margin of a pond, pausing after each step to wiggle first one foot and then the other under the water, stirring up small crustaceans. Still others follow a swimming Red-Breasted Merganser *(Mergus serrator)* at a discreet distance, hoping to catch some leftovers from the fish it disturbs while diving.

The Snowy Egret uses a greater variety of foraging behaviors than any other North American heron, allowing it to adapt to a range of feeding conditions. In addition, groups of feeding Snowy Egrets attract other species, improving the foraging success of all concerned. These techniques have allowed the Snowy Egret to rebound from the depredations of plume hunters earlier this century.

The Snowy Egret is considerably smaller than the Great Egret *(Ardea alba),* often not obvious when the birds are in flight. The important field marks for Snowy Egret identification are the black bill and yellow feet. This is just the reverse of the Great Egret, which has a yellow bill and black feet.

SIZE: Length 24 in. **ADULT BREEDING:** Feb.–July. Sexes similar. Similar to adult nonbreeding, but with short plumes from neck and back. **ADULT NONBREEDING:** Aug.–Jan. Sexes similar. Medium-sized, slender, all-white wading bird. Black legs and yellow feet. Black bill. **STATUS:** Common winter visitor and fairly common migrant. Rare local resident. **(B*)**

GREEN HERON
(Pl. 37)
Butorides virescens

Plate 37. Green Heron: adult.

If, when scanning the reeds that grow beside a freshwater lake or brackish marsh, you should happen to spot a Green Heron, consider yourself fortunate. This, our smallest and shiest heron, shows a dark chestnut neck and lovely bluish green wings and back. When flushed, the Green Heron utters an emphatic, loud "skyow" as it flies away, with crest raised and yellow legs sticking out behind a short tail.

When a Green Heron hunts, it crouches beside the water with neck drawn in. If it spots a minnow, it slowly extends itself, inch by inch, until its neck is suspended horizontally over the water. With a swift grab and a slippery swallow the Green Heron finishes the job, demonstrating its fishing skill.

Green Herons have also been known to bait fish. They place a piece of bait—an insect, flower, or twig—on the water and then snatch an unsuspecting fish once it comes to the surface to investigate.

SIZE: Length 18 in. **IMMATURE:** Sexes similar. Brown streaks on whitish underparts. Duller than adult on wings and back. **ADULT:** Sexes similar. Black crown, dark chestnut head and neck, bluish green wings and back. Yellow legs and feet. **STATUS:** Fairly common migrant and winter visitor. Uncommon local resident. **(B)**

BLACK-CROWNED NIGHT-HERON   *Nycticorax nycticorax*
(Pl. 38)

Watching from the muddy bank of a tidal channel or perched in shrubbery around a city park pond, the hunched shape of the Black-crowned Night-Heron is unmistakable. These white-breasted birds with black caps and backs often doze in midday, preferring to do most of their hunting at night. Black-crowned Night-Herons have an unusually short neck for a heron, which gives them a characteristic round-shouldered look.

Black-crowned Night-Herons breed on every continent except Australia and Antarctica. They are gregarious, nesting and feeding in groups.

Like most herons, Black-crowned Night-Herons do not attain

Plate 38.
Black-crowned
Night-Heron:
adult.

adult plumage until their second year. Immatures are brown, streaky individuals with white spots on their wings. They are sometimes confused with the much rarer American Bittern *(Botaurus lentiginosus)*.

As the sun sets and the tide ebbs, these chunky herons begin to wake up. They are not finicky about what they eat and pursue everything from small fishes to the young of other marshbirds. When Black-crowned Night-Herons fly out to feed, they utter a hoarse, abrupt "quark!," a call frequently heard near the coast at night. Seldom attributed to a heron species, the croak of the Black-crowned Night-Heron announces its presence on the marsh, now that the daytime foragers have departed.

SIZE: Length 25 in. **IMMATURE:** Sexes similar. Broad, brown streaks on white breast, white spots on wings. **ADULT:** Sexes similar. Black head with a white plume, black back, pale gray wings, and whitish breast. Red eye. Black bill. **STATUS:** Fairly common migrant and winter visitor. Uncommon local resident. **(B)**

IBISES
(Threskiornithidae)

The Ibis family is closely related to the herons, storks, and New World vultures. Its Latin name comes from the Greek word *threskeia,* meaning "religious worship" or "sacred," and *ornithos,* meaning "bird."

The ibis has a fossil record going back 60 million years. The Sacred Ibis *(Threskiornis aethiopica)* flourished thousands of years ago in Egypt, where it was worshipped and written about as a religious icon. The bird was mummified and buried in temples with the pharaohs.

Medium-sized dark birds with long legs and a downcurved bill, ibises fly like storks, with outstretched neck and trailing legs.

Ibises frequent freshwater marshes and flooded agricultural fields, searching for insects, newts, earthworms, snails, crustaceans, crayfish, frogs, and fish.

The ibis found on the southern California coast is the White-faced Ibis *(Plegadis chihi).*

WHITE-FACED IBIS

Plegadis chihi

(Pl. 39)

Occasionally during fall and winter, especially after a good rain has soaked the fields and flooded low-lying areas, a few blackish, long-legged birds with downcurved bills can be seen wading along the fringes of shallow freshwater pools.

The shape of the ibis, immortalized by Egyptian artists centuries ago, is striking. Its erect posture and deliberate movements

Plate 39. White-faced Ibis: adult nonbreeding.

are reminiscent of a small, black stork. When foraging, the White-faced Ibis walks slowly forward, often plunging its head beneath the water to retrieve prey.

The White-faced Ibis looks blackish in immature and non-breeding plumage. At this stage, when the bird is most commonly sighted in our area, the white band of feathers that borders the eye and the bare facial skin is missing. As winter turns to spring and the breeding season approaches, the body feathers of the White-faced Ibis turn a deep chestnut color, the wings take on an iridescent green, and the reddish facial skin is, finally, outlined by a white border.

White-faced Ibises nest inland and migrate through portions

of the southern California coast and southward to central Mexico to spend the winter. Their population elsewhere in California has recently expanded, so they are now more likely to be seen passing through our region.

SIZE: Length 23 in. **ADULT BREEDING:** Mar.–Aug. Sexes similar. Dark chestnut body, green iridescence on wings. Reddish facial skin outlined by white border. Red legs. **ADULT NONBREEDING AND IMMATURE:** Sept.–Feb. Sexes similar. Blackish gray body, some greenish gloss on wings. Faint white streaks on neck. Red eye. Pale gray bill. **STATUS:** Rare to uncommon migrant and winter visitor. Small flocks winter at Point Mugu near Oxnard and at Buena Vista and San Elijo Lagoons.

NEW WORLD VULTURES (Cathartidae)

New World vultures are found only in the Western Hemisphere. Although the vultures' soaring behavior would seem to associate them with other birds of prey such as hawks and eagles, they are very different. Recent DNA research concludes that New World vultures are more closely related to the stork family. For this reason, ornithologists now place them in taxonomic order closer to the herons, ibises, and storks.

Vultures have special adaptations for feeding on carrion. They prefer freshly killed carcasses, but putrid ones will do. The bare skin of their unfeathered heads allows them to dip efficiently into the recesses of a messy carcass. Vultures' feet and talons are weak, not suitable for grasping or killing prey, unlike those of hawks and falcons.

All vultures have excellent eyesight, and Turkey Vultures *(Cathartes aura)* have a well-developed sense of smell, too. Unusual in birds, a sense of smell is helpful in locating a decaying animal from a distance.

Vultures take advantage of roadkills for much of their food. They also visit the shoreline to feed on dead marine mammals or fish washed up on the beach.

In flight, vultures show their mastery of the air currents. With wings held in a characteristic shallow V—a dihedral—vultures make slow circles in the sky. Once they set their broad, flat wings, they can cruise for miles on thermals formed by warm air rising. In early morning, before they set out to fly, vultures often hold their wings out to "dry" the way cormorants do.

The Turkey Vulture *(Cathartes aura)* is the only vulture species on the southern California coast.

TURKEY VULTURE *Cathartes aura*
(Pl. 40)

On a winter day when low tide has left a fresh carcass—its ripe smell strong on the ocean air—high and dry on the sandy beach, you can expect to see one or two Turkey Vultures circling overhead within minutes. Seemingly out of nowhere, their black shapes materialize, wings tilting slightly from side to side as they cruise by. Eyeing the carcass, the birds swoop down and land on the beach. Meanwhile, two more appear in the sky. The vulture communication lines are open: when one bird sights prey and descends, others of the flock follow to investigate.

On the sand, the Turkey Vultures hop clumsily toward a dead

Plate 40. Turkey Vulture: adult.

seal. From a distance they look like large, brown lumps, because their featherless heads are so small relative to their bodies. On closer view, the adults have reddish skin on their heads, whereas the heads of the immatures are grayish. Eventually, they start tearing at the carrion with their beaks. As scavengers, Turkey Vultures perform an important recycling function in the natural world.

In fall and winter along the southern California coast, Turkey Vultures gather at communal roosts in tall trees (such as eucalyptus), numbering 70 to 100 birds in certain locations.

SIZE: Length 26 in. **IMMATURE:** Sexes similar. Similar to adult, but grayish skin on head. **ADULT:** Sexes similar. Large, blackish brown soaring bird with pale flight feathers (primaries and secondaries) as viewed from below. Reddish head. **STATUS:** Common migrant and fairly common summer visitor. Uncommon to fairly common local winter visitor.

HAWKS (Accipitridae)

Hawks, also known as raptors or birds of prey, come in a variety of sizes and shapes. They range from large eagles to water-loving Ospreys *(Pandion haliaetus),* from soaring Red-tailed Hawks *(Buteo jamaicensis)* to delicate White-tailed Kites *(Elanus leucurus).* They include fast accipiters that chase birds on the wing, as well as harriers that glide slowly over meadows and marshes.

All hawks have two traits in common: they feed on live prey and they hunt during daylight hours. When hunting, most hawks perch on phone poles or in trees to watch for prey. They also soar or hover in the sky scanning for movement on the ground below. Their eyesight is excellent. Once prey is sighted, hawks swoop down and pounce, catching and killing their victims with sharp, curved talons. When the hunt is over, hawks carry their prey to a nearby perch or eat it on the spot. Roughened patches on the underside of their toes help them carry their prey. Hawks' hooked beaks are suited for plucking and tearing at the flesh of birds, mammals, reptiles, amphibians, fish, and invertebrates.

Along the southern California coast, the following hawks can be observed: Osprey *(Pandion haliaetus)*, White-tailed Kite *(Elanus leucurus)*, Northern Harrier *(Circus cyaneus)*, Red-shouldered Hawk *(Buteo lineatus)*, and Red-tailed Hawk *(B. jamaicensis)*.

Two species of accipiter, Cooper's Hawk *(Accipiter cooperii)* and Sharp-shinned Hawk *(A. striatus)*, while common inland, are not consistently found in coastal habitats and are not included here.

OSPREY   *Pandion haliaetus*

(Pl. 41)

In fall or winter at estuaries along the coast or at freshwater marshes just inland, a long-winged brown-and-white bird circles above the water. It calls "kewp, kewp, kewp" in a high-pitched whistle. Sometimes the Osprey hovers high over the water, tail spread, feet dangling. Spotting a fish, it dives with feet forward, wings held above the back. The bird makes a huge splash in the water from the force of the plunge. When it rises, a fish wriggles in its talons. Then the Osprey heads in lumbering flight to a dead snag, at the same time rearranging the fish so that it faces forward to provide less air resistance. Here, the Osprey will devour its catch.

Plate 41. Osprey: adult (Peter LaTourrette).

The Osprey's fish-eating habits make it unique among the hawks. It is sometimes classified in its own family due to special modifications: the Osprey's undertoes are covered with little spines (spicules) that help hold onto slithering fish; its outer toe reverses so that it can grasp prey with two toes in front and two in back.

In a recent study to determine fall migratory movements of Ospreys, satellite telemetry was used to track Western Ospreys that had nested along the Columbia River between Oregon and Washington. In migration, they moved through California, and most wintered in Mexico, with a few going to El Salvador and Honduras. Western birds traveled a shorter distance than either the Midwestern or Eastern Ospreys in the study group. In addition, the females from each nesting population wintered farther south than the males.

The Osprey is one of the most widely distributed of all bird species, being found in Europe, Asia, North Africa, South China, the East Indies, Australia, and the southwest Pacific.

SIZE: Length 23 in. **ADULT:** Sexes similar. Long, crook-winged silhouette in flight. White head with dark line through eye. Brown back, white underparts. Barred tail. **STATUS:** Uncommon migrant and winter visitor. Rare summer visitor.

WHITE-TAILED KITE   *Elanus leucurus*
(Pl. 42)

A beautiful, apparently all-white hawk hovers with rapid wing beats high over marshes, grasslands, or, occasionally, highway medians. The hawk's whiteness and its conspicuous flight behavior attract the attention of passersby.

By hovering with beating wings raised, the White-tailed Kite holds itself stationary in the air, scanning the ground—an adaptation for grassland areas where few trees are available for lookouts. Once the kite spots a mouse or vole, the bird drops down feet first, wings held high above its back, and captures the rodent. The White-tailed Kite is more delicate and graceful than other hawks. With its shorter legs and weaker feet, it seeks smaller food than more aggressive raptors.

Ornithologists believe that White-tailed Kite abundance is tied to that of its primary prey in California, a tiny rodent, the

Plate 42. White-tailed Kite: adult.

California Vole *(Microtus californicus)*. When the voles become scarce due to lack of food during dry conditions, the birds have a difficult time finding enough prey to sustain themselves. As a result, the population of White-tailed Kites in southern California is subject to fluctuating long-term cycles. Before the 1930s, this species was considered rare. In the mid-1970s, the population peaked, before taking a dive during the drought of the mid-1980s through the early 1990s.

At present, White-tailed Kites seem to be holding their own in southern California, although numbers are still subject to prey population crashes. The conversion of ungrazed grassland, which they prefer for foraging and nesting, to urbanization and agriculture continues to be a problem.

SIZE: Length 15 in. **ADULT:** Sexes similar. Pale gray above, white below, white head, white tail. Black patch on shoulder visible from above and below. **STATUS:** Uncommon to fairly common resident. **(B)**

NORTHERN HARRIER   *Circus cyaneus*
(Pl. 43)
Where salt marshes create lowlands along our southern coast, and at grassy bluffs from Santa Barbara County northward, a slender-bodied, long-tailed hawk occasionally appears, flying just inches

Plate 43. Northern Harrier: adult female (Peter LaTourrette).

above the low vegetation. With wings held slightly upward in a V, the hawk glides slowly back and forth across open country looking for prey. At one point, it flashes a distinctive white rump patch as it turns sharply to investigate a wiggle in the grass.

The Northern Harrier, formerly called the Marsh Hawk, is the only U.S. species of a group of hawks more common in the Old World. The Northern Harrier hunts by flying low over open fields, searching for mice, rats, frogs, and small birds. Spying prey, it interrupts its flight with a swift, agile pounce. When perching to survey its surroundings, it is apt to choose a low stump or even the ground. In addition, the Northern Harrier has owl-like facial discs that help it locate prey by sound, as well as by sight.

Many Northern Harriers migrate long distances; some nest in Alaska and winter in northern South America. Our southern California birds are chiefly winter visitors, but a few pairs still nest at the Tijuana River valley and at locations north of Point Conception.

SIZE: Length 18 in. **ADULT MALE:** Pale gray above, gray facial discs, white below, white rump. **ADULT FEMALE AND IMMATURE:** Brown above, brown-streaked below, pale brown facial discs, white rump. **STATUS:** Uncommon migrant and uncommon to fairly common winter visitor. Rare local resident. **(B*)**

RED-SHOULDERED HAWK   *Buteo lineatus*

(Pl. 44)

On clear February days when the sun shines after a rain, a loud, ringing cry comes from two hawks circling together in the sky, "kee-YER, kee-YER, kee-YER, kee-YER!" Looking up, you can see the hawks have narrow horizontal bands on their tails and pale, translucent "windows" toward the end of their wings.

These are Red-shouldered Hawks, one of the most vocal species of the hawk family, especially during courtship flights over suburban streamside corridors and coastal sage scrub. Red-shouldered Hawks have adapted well to humans and frequently build their nests in tall pines or eucalyptus trees in parks, or in palm trees along city streets.

Plate 44. Red-shouldered Hawk: adult.

Both Red-shouldered Hawks and Red-tailed Hawks *(B. jamaicensis)* belong to the genus *Buteo*—closely related hawks that soar high in the sky. Red-shouldered Hawks use a characteristic "flap, flap, flap, sail" rhythm in flight, beating their wings quickly several times, then gliding. When hunting, Red-shouldereds sit scrutinizing the ground from a nearby perch.

The California population of Red-shouldered Hawks is rather sedentary and does not migrate. It is also isolated, being separated from its eastern relatives by the Rockies and the Great Plains, where the species is absent.

SIZE: Length 17 in. **ADULT:** Sexes similar. Brownish upperparts, rufous barring on underparts. Extensive white spots on back. Narrow black-and-white bands on tail. **STATUS:** Fairly common resident. **(B)**

RED-TAILED HAWK   *Buteo jamaicensis*
(Pl. 45)

Plate 45.
Red-tailed
Hawk: adult.

Whether sitting in traffic on a Los Angeles freeway or hiking on a lonely coastal trail, a birder seeing a hawk soaring in wide circles against a blue sky is likely to call it a Red-tailed Hawk. This hawk holds itself aloft by floating on the air currents, seldom flapping its broad wings. It uses its reddish brown tail as a rudder, tilting first to one side then another.

In spring, the Red-tailed Hawk may be joined by its mate in a wonderful display flight. Together they parachute downward, side by side, with legs dangling. Occasionally, one of the birds gives a single, descending scream "KREEeeee-ee-ee-ee"—a wild, fierce cry. Both birds soon alight on a favorite perch near their nest tree.

Adult Red-tailed Hawks are easiest to identify in flight, with their reddish brown tails. For young birds, which do not have reddish brown tails, the single best identifying mark is the dark strip along the leading edge of the wing or "shoulder" in flight, as seen from below.

Red-tailed Hawks are one of the most widespread birds of prey in North America. In southern California, they are year-round residents, with the population augmented in winter months by individuals from northerly regions. Red-taileds prefer more open sites than Red-shouldered Hawks *(Buteo lineatus)*. Look for them in the countryside, often perched on utility poles along the road.

SIZE: Length 19 in. **IMMATURE:** Sexes similar. Similar to adult, but tail has narrow black-and-white bands, lacks reddish brown color. **ADULT:** Sexes similar. Brown upperparts, dark head, white belly crossed by thick, dark band. Underparts in flight show dark mark along leading edge of wing (patagial area), reddish brown tail. **STATUS:** Common migrant and winter visitor. Fairly common resident. **(B)**

FALCONS (Falconidae)

Falcons are a group of raptors designed to chase and catch their prey in flight. They are fleet as the wind and pursue other birds.

Falcons have several specialized structures suiting them for high-speed pursuit and swift killing. Long, pointed wings that extend to the tip of their tail when perched help streamline falcons' flight. Their bills are equipped with a "tooth" in the top mandible and a notch in the lower one, enabling the falcon to bite through the victim's neck, instantly breaking the vertebrae. Falcon nostrils have a bony tubercle that blocks out excess oxygen when the bird is diving fast. And their heavy brows and dark "mustaches" reduce the glare of the exposed cliffs and treeless shores where they do most of their hunting.

Three species of falcon frequent the southern California coast: the American Kestrel *(Falco sparverius)*, Merlin *(F. columbarius)*, and Peregrine Falcon *(F. peregrinus)*.

AMERICAN KESTREL   *Falco sparverius*
(Pl. 46)
A small falcon hovers, using rapid wing beats as it holds itself high above a field or roadside ditch. It faces into the wind and scans

Plate 46. American Kestrel: adult male.

below for prey, spreading a rufous tail. Then, the falcon flies to a utility wire where it alights, giving a characteristic tail-bobbing motion, as though to regain balance on a tightrope, and uttering a high "ki-ki-ki-ki" to warn off interlopers.

The American Kestrel, formerly known as the Sparrow Hawk, rarely chases birds in midair the way other falcons do. Instead, it hovers like a kite, then swoops down to pounce on insects, mice, and lizards.

Throughout North America, American Kestrels are the most widespread falcon. They live in open areas with short, grassy vegetation—tolerating human-made habitats such as agricultural fields and suburban parks. Because they are cavity nesters, American Kestrel numbers are limited by the availability of nesting holes. With the discovery that these birds will successfully use artificial nest boxes, biologists and nature lovers have been able to lure the species back to habitats it formerly occupied.

SIZE: Length 9 in. **ADULT MALE:** Upperparts reddish brown, blue gray wings, two black facial stripes. **ADULT FEMALE:** Similar to adult male but lacks blue gray wings. **STATUS:** Common migrant and winter visitor. Fairly common local resident. **(B)**

MERLIN   *Falco columbarius*

(Pl. 47)

On a January day at the salt marsh, a small, dark falcon about the size of an American Kestrel *(Falco sparverius)*, streaks low over the brown pickleweed. By the time an observer can focus the binoculars, the falcon dashes away, its flight marked by quick, deep wing beats. The falcon heads for the mouth of the estuary, bent on picking out one of the small shorebirds feeding on the mudflats. In the distance, the silver-winged flocks rise and swirl in the sky, their warning calls coming too late. The sickly, the inexperienced, and the careless among the small shorebirds are fair game for the Merlin: it will not go home hungry this morning.

Merlins breed throughout northern North America, Europe, and Asia; their name comes from the Old French term for this species, *esmerillon.* Along the southern California coast, watch for a hawk that is a darker, slightly larger version of the American

Plate 47. Merlin:
adult female
(Peter LaTourrette).

Kestrel, with heavier streaking below. In fall and winter, wherever small shorebirds—such as Least Sandpipers *(Calidris minutilla)* and Dunlin *(C. alpina)*—congregate, the Merlin may occasionally appear, hugging the ground in low flight as it attempts to sneak up on the feeding flocks.

SIZE: Length 10 in. **ADULT MALE:** Faint "mustache" stripe. Dark, blue-gray upperparts, rufous-streaked breast, dark tail with pale bands. **ADULT FEMALE:** Similar to male, but browner upperparts, brown-streaked breast. **STATUS:** Uncommon to rare migrant and winter visitor.

PEREGRINE FALCON
(Pl. 48)

Falco peregrinus

Along the bluffs at Point Vicente, at the tidal mudflats of Goleta Beach, or at the enormous rock at Morro Bay, a gray shadow comes hurtling down out of nowhere. At Point Vicente, the pigeons whirl and scatter in terror. At Goleta Beach, the gulls and terns rise in screaming disarray. At Morro Rock, the shorebirds fling themselves into the sky. A Peregrine Falcon is hunting.

Peregrine Falcons fly at high speed with deep, fast wing beats. When they see prey, they fold their wings slightly and dive down-

Plate 48. Peregrine Falcon: adult.

ward upon their flying quarry. This downward dive, called a stoop, has been clocked at 89 to 174 miles per hour, the fastest speed of any bird.

The Peregrine hits its target with all talons extended, sending a shower of feathers out of the sky. As the victim tumbles earthward, the Peregrine may snatch it on a return pass, or let the bird fall to the ground. Often, the falcon dines on the spot, regardless of its surroundings. Peregrine Falcons feed on everything from ducks to songbirds. Rock Pigeons *(Columba livia)* (feral pigeons) are a favorite.

If the Peregrine passes close enough in flight or decides to perch on a cliff face within binocular view, notice its dark "mustache" and dark head: together they give a helmeted look to the bird.

Peregrine Falcon populations in North America suffered from DDT poisoning and eggshell thinning in the 1950s and 1960s. The near extinction of this magnificent falcon was an impetus to environmentalists to save the species. In 1970, the bird was listed by the federal government as endangered. After an intense campaign on the part of many organizations, Peregrine Falcon populations began to recover. With the banning of DDT and the success of hacking (captive-raised birds are reintroduced into the wild while being fed until they become independent), the Peregrine Falcon appears to be doing well. Although the bird was delisted (no longer considered endangered) by the federal government in 1999, the subspecies known as the American Peregrine Falcon *(F. p. anatum)* is still on the California state endangered list.

Peregrine Falcons (the word "peregrine" means "a wanderer from a foreign land") are aptly named: they travel great distances in migration. Most of the Peregrines we see in southern California are wintering from Alaskan breeding areas; however, a small but increasing number of Peregrines are permanent residents here: a few have taken up nesting atop tall buildings in downtown Los Angeles, while others choose cliff sites along the coast.

SIZE: Length 16 in. **IMMATURE:** Sexes similar. Similar to adult, but with dark brown upperparts. **ADULT:** Sexes similar. Dark head with dark "mustache" gives helmet effect. Slate gray upperparts, barred underparts. **STATUS:** Uncommon to rare migrant and winter visitor. Rare local resident. **(B*, SE)**

RAILS, GALLINULES, and COOTS (Rallidae)

The phrase "thin as a rail" refers to the rail's ability to squeeze through tiny passages between dense marsh plants. Rails can flatten their feathers tightly to their bodies, sliding secretively through cattails and tules without a trace.

Rails, gallinules, and coots comprise a group of short-winged, stubby-tailed birds that frequent marshes. Rails are solitary and hard to see; coots and gallinules are much more visible. All members of this family, the Rallidae, can swim and dive, but coots are the most agile in the water. All have long, thin toes for walking over lilypads and submerged reeds, but coots have lobes along each toe to facilitate swimming.

Many people have never seen a rail. By waiting patiently at the edge of a marsh at dawn or dusk, birders hope to see or hear these reclusive birds: rails call loudly as they sneak through hidden, watery channels, communicating with each other by vocalizations.

Rails have declined in coastal southern California. The draining and urbanizing of coastal wetlands reduced their habitat drastically in the latter part of the twentieth century. Even so, rails persist and can be seen, particularly during winter months, in areas where marshes have been preserved or restored.

The southern California coast has three species of regularly occurring rails: the Clapper Rail *(Rallus longirostris)*, the Virginia Rail *(R. limicola)*, and Sora *(Porzana carolina)*. The gallinules are represented by the Common Moorhen *(Gallinula chloropus)*, and the coots by the American Coot *(Fulica americana)*.

CLAPPER RAIL   *Rallus longirostris*
(Pl. 49)

As dawn steals across the salt marsh at Tijuana Slough or Upper Newport Bay, a chickenlike bird picks its way cautiously along the exposed margins of the mud. It stays near the edge of the pickleweed, jabbing at morsels on the ground with a slender, slightly drooping orange bill. The bird's long neck and bill are out of proportion to its short tail. As it walks, the rail flicks its tail nervously, flashing white undertail coverts.

This is the Clapper Rail, a shy species that lives only in coastal salt marshes. More frequently heard than seen, the Clapper Rail utters a loud series of "cac-cac-cac-cac" notes that begin slowly, then gradually run together. The sound is like two hands clapping or two pieces of wood hitting against each other, and is usually given by paired birds during breeding season.

At extreme high tides, such as those occurring in late fall and early winter, Clapper Rails are flooded out of their hiding places and forced to the margins of the salt marsh, where they may be spotted foraging in the open. They feed on crabs, spiders, and snails by gleaning the surface of the mud.

Plate 49. Clapper Rail: adult (David Koeppel).

Several subspecies of Clapper Rail inhabit California, but the Light-footed Clapper Rail *(R. l. levipes)* is the one that breeds in coastal southern California salt marshes. Given that salt marshes are scarce in our region and have been degraded over the past several decades, it is not surprising that Light-footed Clapper Rails have been listed as endangered by both the federal and state governments since the early 1970s. In addition to loss of habitat, Light-footed Clapper Rails have suffered from an explosion of predators such as nonnative Red Foxes *(Vulpes vulpes)*, Raccoons *(Procyon lotor)*, Opossums *(Didelphis virginiana)*, and stray pets, all of which prey on rail eggs and young.

At present, Light-footed Clapper Rails are holding their own in the larger salt marshes where their populations remain viable, among them Tijuana Slough, Upper Newport Bay, and Seal

Beach. Thanks to intensive management of the birds by government agencies and dedicated researchers, Light-footed Clapper Rails have slowly recovered, although smaller marshes harbor very few. In 1996, an estimated total of 325 pairs of Clapper Rails inhabited coastal southern California salt marshes. Upper Newport Bay shelters a remarkable 50 percent of the population.

Since 2001, a captive breeding program operating at the Chula Vista Nature Center, adjacent to Sweetwater Marsh National Wildlife Refuge near San Diego Bay. Researchers plan to release these captive-bred and wild-caught birds in places such as Carpinteria Salt Marsh where suitable habitat exists, but where the rail population has declined sharply. Currently, it is estimated that fewer than 600 Light-footed Clapper Rails exist in the wild.

SIZE: Length 14.5 in. **ADULT:** Sexes similar. Olive brown with dark streaks above, gray face, deep cinnamon breast, flanks and back darker. Long orange bill. **STATUS:** Uncommon local resident south of Point Conception. **(B*, FE, subsp. SE)**

VIRGINIA RAIL   *Rallus limicola*
(Pl. 50)

From the tules and cattails of freshwater marshes or lakes, a metallic "kid-ick, kid-ick, kid-ick" grates on the ears. Other marsh dwellers such as the Red-winged Blackbird *(Agelaius phoeniceus)*

Plate 50. Virginia Rail: adult.

and the Marsh Wren *(Cistothorus palustris)* add their spring songs to the dawn chorus, but what is this bird that sounds like a creaking wagon wheel one minute and a grunting pig the next? This is the Virginia Rail, which gives a variety of odd calls. In spring, grunting duets are given by mated pairs.

At dawn or dusk, a Virginia Rail sometimes forages near the edges of its reedy habitat. If you are lucky enough to be standing close by, you notice that a small, chunky shape with a cinnamon red breast has emerged from the tules. The bird steps carefully over the mudflats as it walks with short, upcocked tail. It moves slowly, poking the ground with a long, reddish bill for earthworms, snails, and larval insects.

In coastal southern California, Virginia Rails are often found in salt marshes in winter, where they might be confused with Clapper Rails *(R. longirostris)*. Virginia Rails are much smaller than Clapper Rails and their coloring is more vivid.

SIZE: Length 9.5 in. **ADULT:** Sexes similar. Gray face, rich cinnamon red breast, black-and-white barring on flanks. Reddish bill. **STATUS:** Uncommon to fairly common migrant and winter visitor. Uncommon local resident. **(B*)**

SORA   *Porzana carolina*
(Pl. 51)

A plump-bodied bird with a short, yellow bill skirts the edges of the reeds at the San Joaquin Wildlife Sanctuary, Lake Los Carneros, Oso Flaco Lake, and other freshwater marshes. As it threads its way through the tall stems, the bird flicks its white undertail coverts. If it comes to an inlet too deep to wade across, it swims, then clambers up onto the next patch of decaying tule stems and resumes searching for seeds and insects.

The Sora, the most common of North American rails, is also the easiest to see. Smaller than the Virginia Rail *(Rallus limicola)* and less secretive, the Sora rewards a patient birder by occasionally showing itself while exploring the fringes of dense marsh plants around bodies of water. The loud, downward whinnying call of the Sora, often heard around marshes, is a helpful clue to its presence.

Soras would appear to be weak fliers, but they can migrate long distances. Some of the Soras that overwinter in southern California have migrated here from northern Canada.

Plate 51.
Sora: adult
nonbreeding.

SIZE: Length 8.75 in. **ADULT BREEDING:** Mar.–Aug. Sexes similar. Brown upperparts, black-and-white barred flanks, black patch on face and throat. Small, yellow bill. **ADULT NONBREEDING:** Aug.–Feb. Sexes similar. Similar to adult breeding, but reduced black on face. **STATUS:** Uncommon to fairly common migrant and winter visitor. Rare local resident. **(B*)**

COMMON MOORHEN *Gallinula chloropus*
(Pl. 52)

Hidden in the backwater channels of lakes and ponds, a bird that at first glance appears to be a coot swims along, pumping its head forward with each stroke. Dark like a coot, but smaller and more

Plate 52.
Common
Moorhen:
adult
breeding.

delicate, the Common Moorhen's striking red-and-yellow bill sets it apart. The Common Moorhen is browner than the coot above and has a white horizontal stripe along its side.

The Common Moorhen, formerly Common Gallinule, is intermediate between a rail and a coot. It swims in the corners of the marsh, not flaunting its presence on open water the way a coot does. It is able to walk along the water's surface on submerged vegetation, balancing on long, slender toes.

In coastal southern California, Common Moorhens are an uncommon sight, observed mostly during winter months.

SIZE: Length 14 in. **ADULT BREEDING:** Feb.–Sept. Sexes similar. Grayish black head, neck, and underparts, brown back. White stripe along flank, white undertail coverts. Red frontal shield and base of bill, yellow bill tip. **ADULT NONBREEDING:** Sept.–Feb. Sexes similar. Similar to adult breeding, but lacks red shield. Dusky bill. **STATUS:** Uncommon migrant and winter visitor. Rare local resident. (B*)

AMERICAN COOT   *Fulica americana*
(Pl. 53)

The rails and the Common Moorhen *(Gallinula chloropus)* are shy and solitary, but their relative, the American Coot, is gregarious and easy to see. Sometimes known as Mud Hens, these black birds with white shields on their foreheads float on open lakes and brackish lagoons anywhere along the coast. From a distance, they might be ducks, except they pump their heads back and forth when swimming.

American Coots have adapted well to human-made habitats. They like artificial ponds in parks or other urban surroundings, where they graze on the planted grassy slopes. Sometimes American Coots wander in large groups onto golf courses, so long as they can escape to a nearby lake when threatened.

On land, these birds waddle in a hunched shape, but in the water they are skilled divers. American Coots' lobed toes allow them to dive to 25 feet in search of small fish, snails, or pond vegetation. They are not picky eaters, but will steal food brought up by ducks, and in turn are robbed of their own food by American Wigeons *(Anas americana)* and Gadwalls *(A. strepera)*.

During mating season, American Coots splash about and

Plate 53.
American
Coot: adult.

show off. They charge across the surface of the water at rivals, occasionally attacking an intruder with their claws. They whistle, croak, and grunt. A low, staccato "kuk-kuk-kuk-kuk" is typical, each syllable uttered in time with the bird's bobbing head as it swims along.

SIZE: Length 15.5 in. **ADULT:** Sexes similar. All dark, slate gray body, black head. White frontal shield and bill. **STATUS:** Common migrant and winter visitor. Fairly common resident. **(B)**

PLOVERS (Charadriidae)

On a beach walk, a birder notices several kinds of shorebirds running along the wet sand or exploring the piles of kelp tossed up by the tide. Although they may be different sizes and shapes, the majority of the birds are members of two families: the plovers and the sandpipers.

Identifying plovers is easy because of their distinctive shape: short bill, round head, short neck, compact body. They have comparatively large eyes for their small, rounded heads, giving them a wide-eyed, innocent look.

Plovers' short bills adapt them for feeding in a unique fashion. They hunt best on expanses of level ground, where they first see prey, then run to snatch it. Between forays, they pause, stand erect, then dart over and grab the next worm or insect. Occasionally, a plover pats the surface of the wet sand with its toes to disturb prey.

Coastal southern California hosts four species of plover: Black-bellied Plover *(Pluvialis squatarola)*, Snowy Plover *(Charadrius alexandrinus)*, Semipalmated Plover *(C. semipalmatus)*, and Killdeer *(C. vociferus).*

BLACK-BELLIED PLOVER   *Pluvialis squatarola*
(Pl. 54)

On a winter beach near the high-tide line, a group of medium-sized shorebirds dozes with their heads tucked into their back feathers. Their upperparts are speckled with black and white, like a tweed coat. Until they awake and show their bills, they are difficult to identify.

Frightened by a dog galloping across the sand, the birds stir, then leap into the air with flashing wings. As they fly away, the plovers show a patch of black feathers near the body on the underwing. They give a wild, questioning call "tlee-oo-ee?," the middle note lower than the first and last.

Black-bellied Plovers may look drab in nonbreeding plumage, but they undergo a dramatic molt in spring, donning the black bellies for which they are named. During their stay on southern California beaches, they spread out to forage singly, their large

Plate 54. Black-bellied Plover: adult nonbreeding.

eyes enabling them to see prey even at night. When roosting and flying, however, they gather in flocks.

Far-ranging, strong-flying migrants, Black-bellied Plovers are the largest of the four plovers that frequent coastal southern California. They travel northward in spring to nest on the tussocks and gravel of the high Arctic, returning in fall to winter along the Pacific coast as far south as Chile.

SIZE: Length 11.5 in. **ADULT BREEDING:** Apr.–Sept. Sexes similar. Black-and-white speckled back, black chest and belly outlined with white on head and upper chest. Black near body on underwings (axillaries). **ADULT NONBREEDING:** Aug.–Apr. Sexes similar. Black-and-white speckled back, white underparts, black near body on underwing (axillaries), pale eyebrow. **STATUS:** Common migrant and winter visitor. Uncommon summer visitor.

SNOWY PLOVER   *Charadrius alexandrinus*
(Pl. 55)

The wet sand where the waves rush up and back on the shore is the scene of the greatest shorebird activity, as many species sift through morsels uncovered by the retreating waves. But the dry sand between the lines of washed-up kelp and the dunes is much less attractive to birds—less attractive to all but the diminutive Snowy Plover.

The Snowy Plover dresses in pale gray brown with an incomplete breastband. Its pattern of a light back and darker head markings blends with bits of shell and stone strewn on the pale sand. When the Snowy Plover stands motionless on the beach, it is all but invisible.

Snowy Plovers, unlike their relations the Black-bellied and Semipalmated Plovers *(Pluvialis squatarola* and *C. semipalmatus),* do not migrate to Arctic regions. Although many leave to nest on alkali lakes in the interior West, others remain on the beaches of southern California year-round. And that has been the bird's undoing.

For southern California's famous beaches attract crowds of sunbathers and swimmers. During summer days, beach sand is groomed and raked, making the plovers more obvious to predators such as the Red Fox *(Vulpes vulpes),* American Crow *(Corvus brachyrhynchos),* Common Raven *(C. corax)*, feral cats, and unleashed pets.

Plate 55. Snowy Plover: adult nonbreeding.

Southern California's Snowy Plovers belong to the Western Snowy Plover *(C. a. nivosus)* complex, whose Pacific coast population ranges from Washington State south to Baja California. Since 1970, Pacific coastal bird numbers have declined, due to loss of their nesting habitat and human disturbance.

In 1993, the federal government declared the Pacific coastal population of the Western Snowy Plover threatened. In 1999, areas of critical habitat were designated, so the birds' nesting beaches would receive monitoring and protection. High public interest and the success of monitoring programs have begun to reverse Snowy Plover breeding colony declines in certain areas along the coast. In addition, Snowy Plovers gain some reproductive advantage when they choose to nest within California Least Tern *(Sterna antillarum browni)* colonies, this tern being listed for some time as endangered and, typically, protected by a fenced enclosure (see also page 47).

Recently, Snowy Plover nesting colonies have expanded at Sands Beach at Coal Oil Point in Goleta, and at Hollywood-by-the-Sea Beach and Ormond Beach near Port Hueneme. Other colonies exist at Tijuana Slough, South San Diego Bay, lagoons north of San Diego, Bolsa Chica, the Santa Clara River estuary, Vandenberg Air Force Base, Pismo Beach, and Morro Bay.

SIZE: Length 6.25 in. **ADULT BREEDING:** Feb.–Aug. Sexes similar. Pale gray upperparts, white underparts. Dark patches on head as follows: above white forecrown and ear coverts. Incomplete breast band, resulting in two dark shoulder patches. Black bill. **ADULT NONBREEDING:** Sept.–Feb. Sexes similar. All dark markings are gray brown. **STATUS:** Fairly common, but local migrant and winter visitor. Uncommon local resident. **(B*, FT)**

SEMIPALMATED PLOVER   *Charadrius semipalmatus*
(Pl. 56)

On December days when extreme low tides pull the waves far out, leaving a long expanse of wet sand on the empty beach, a small, brown-backed shorebird runs quickly, then pauses, head up, to survey its surroundings. The brown of the bird's upperparts blends into the brown of the dark sand. When it moves again, it shows yellowish legs and a single black breastband across the upper chest—like a miniature Killdeer *(C. vociferus)* with one band, not two.

Plate 56. Semi-palmated Plover: adult nonbreeding.

The Semipalmated Plover, named for the partial webbing between its three front toes (not visible in the field), is easily missed if it stands perfectly still on estuary mudflats or the wet sand of open beaches. Often in the company of Least Sandpipers *(Calidris minutilla)* and Sanderlings *(C. alba),* Semipalmated Plovers are much less active. And they do not feed in close flocks like the Least Sandpipers, but spread out on the strand, each rather solitary in its own feeding space.

When startled, Semipalmated Plovers are quick to come together as a flock and fly away. Their rising two-note call "chu-wee?" is a familiar background to fall and winter shorebirding expeditions.

Semipalmated Plovers breed on Arctic tundra and winter on coasts from the southern United States south through most of South America.

SIZE: Length 7.25 in. **ADULT BREEDING:** Mar.–Sept. Sexes similar. Brown upperparts, white underparts, single dark breastband. Dark patches on head as follows: on forecrown and from bill through eye. Yellowish legs. Yellowish base of bill. **ADULT NON-BREEDING:** Sept.–Mar. Sexes similar. All dark markings paler brown. Legs and bill base paler. **STATUS:** Common migrant and fairly common winter visitor. Uncommon summer visitor.

KILLDEER
Charadrius vociferus

(Pl. 57)

At almost any patch of open space—a gravel road, a playing field, a golf course, or a stretch of bare dirt—a birder is likely to hear a loud, ringing "kill-DEE, kill-DEE, kill-DEE!" The vociferous Killdeer circles overhead, landing not too far away. After it lands,

Plate 57. Killdeer: adult.

it teeters its body and bobs its head, showing two black bands across its chest and a rusty orange rump.

Probably the most widely distributed and familiar of North America's shorebirds, Killdeer are sometimes found far from the coast. They are usually never far from water, however, for they feed on a variety of insects gleaned from fields and lawns. Killdeer have managed to survive in southern California landscapes by adapting well to human-made habitats.

The Killdeer is one of the few shorebirds that remains in Southern California in spring to breed along our coast. Its nest, a simple depression that holds the eggs, is lined with pebbles. An adult sitting on the nest is well camouflaged, and subtle markings on the eggs and young make them nearly impossible to detect.

If you flush a Killdeer from its nest, make good notes of the location, for your attention will immediately be drawn elsewhere. An apparently injured, pathetic Killdeer just ahead in the middle of the road will flop and feint, wings drooping, tail fanned to show the orange rump: all of this to lure you away from the nest. If you follow the bird, it will hop just out of reach, demonstrating the crippled bird act all the way. Both male and female Killdeer use this behavior to distract predators.

SIZE: Length 10.5 in. **ADULT:** Sexes similar. Brown upperparts, white underparts, two black breastbands. Black on head as follows: above white forehead and from bill beneath eye. Rusty uppertail coverts and rump. **STATUS:** Common migrant and visitor. Fairly common resident. **(B)**

OYSTERCATCHERS
(Haematopodidae)

The Latin name for this striking shorebird comes from the Greek *haimatopous,* meaning "blood foot." Indeed, the oystercatcher's pink feet and legs and bright red bill are so eye-catching as to be visible at great distances.

Oystercatchers live on rocky coasts, where they feed on the shellfish bounty exposed at low tide. The mussels, clams, chitons, limpets, and barnacles that cling to the wave-washed rocks of the outer beach are favorite foods. Few other birds are as well equipped to forage up and down the slippery rocks. Small structures on the oystercatcher's partially webbed feet allow it to cling to wet surfaces.

The bird wields its bill like an oyster knife, a tool to deal with hard-shelled mollusks. The bill is laterally compressed (flattened) and has a chisel-like tip. Prizing mussels or clams from their berths, the oystercatcher inserts its bill into the animals, severing the muscle holding the two halves of the shell together. Once the shell is opened, the bird helps itself to a nourishing meal. In a year, an oystercatcher can eat more than 100 pounds of mussel meat.

The Black Oystercatcher *(Haematopus bachmani)* is the species inhabiting southern California's shores.

BLACK OYSTERCATCHER   *Haematopus bachmani*
(Pl. 58)

By carefully scanning the ledges and platforms of the rocks at Point Fermin, or the jetties at Playa del Rey or Ventura Harbor, a

Plate 58. Black Oystercatcher: adult.

birder hopes to detect a pair of pink legs and a reddish bill—the only easily visible parts of this dark bird as it picks its way among the dark, wave-washed rocks.

Black Oystercatchers do not hide. They proclaim themselves with loud, harsh "wheeep, wheeep!" cries as they fly from one rock to another. Of the rocky coast shorebird species inhabiting southern California—Surfbird *(Aphriza virgata),* Wandering Tattler *(Heteroscelus incanus),* and Black and Ruddy Turnstones *(Arenaria melanocephala* and *A. interpres)*—the Black Oystercatcher is the only one to nest here. In spring, they engage in territorial squabbles, setting up nesting sites on offshore islets or deserted jetties.

Although Black Oystercatchers are resident on rocky portions of the Pacific coast from southern Alaska to Baja California, it is always an achievement to find one south of Point Conception.

SIZE: Length 17.5 in. **ADULT:** Sexes similar. All-black plumage. Pinkish legs and feet. Long, orange red bill. **STATUS:** Fairly common resident north of Point Conception, and uncommon local resident south of Point Conception. Uncommon to rare migrant and winter visitor elsewhere. **(B)**

STILTS and AVOCETS (Recurvirostridae)

With their long, gangling legs and spectacular black-and-white plumage, stilts and avocets are conspicuous.

Stilts and avocets frequent coastal estuaries and shallow wetlands, seldom venturing onto the open beach. They hunt for food in either brackish or freshwater ponds, eating aquatic bugs, beetles, dragonfly nymphs, and the like. Their bills are long and thin: the avocet's curves upward, and the stilt's is straight.

Both the Black-necked Stilt *(Himantopus mexicanus)* and the American Avocet *(Recurvirostra americana)* frequent the southern California coast.

BLACK-NECKED STILT   *Himantopus mexicanus*
(Pl. 59)

As spring turns to summer at San Joaquin Wildlife Sanctuary, a bird with a black back, white underparts, and long, dangling, pinkish red legs flies over one of the shallow impoundments. It keeps up a constant, nerve-wracking "yip-yip-yip-yip" call. Once landed along the dike several yards ahead, the bird is joined by another. The second bird starts staggering around, then collapses as if badly injured. Through binoculars, a birder sees two pipe-stem-

Plate 59. Black-necked Stilt: adult female.

legged fluff balls wobbling down the gravel pathway some distance in front. What's going on here?

These Black-necked Stilts, intent upon protecting their newly hatched young, are vociferous and conspicuous in defense of their territory. The female is feigning injury to lead interlopers away from the juvenile stilts.

Outside of the breeding season, Black-necked Stilts are much quieter. They forage for floating aquatic bugs and beetles in shallow pools, adroitly bending their stalklike legs to dart at prey.

Black-necked Stilts are more likely to be found near freshwater habitats than their American Avocet *(Recurvirostra americana)* cousins.

SIZE: Length 14 in. **ADULT MALE:** Black upperparts, white underparts. Long pinkish red legs. Thin black bill. **ADULT FEMALE:** Similar to adult male, but back tinged with dark brown. **STATUS:** Common migrant and fairly common winter visitor. Fairly common local summer resident from Santa Maria River estuary south. **(B)**

AMERICAN AVOCET   *Recurvirostra americana*
(Pl. 60)

On a cold, overcast day when a coastal estuary is bathed in gray, a group of tall shorebirds with black-and-white-patterned backs flies in. The American Avocets eventually alight, breaking their landing by touching down with long, blue gray legs. In contrast to

Plate 60. American Avocet: adult male breeding.

the somber browns of the other shorebirds feeding nearby, these black-and-white beauties with their graceful upcurved bills command attention.

Come spring, the heads and necks of male and female American Avocets are suffused with an apricot cinnamon blush in breeding plumage. Avocets show sexual differences in bill shape: the male's bill is longer and less curved, while the female's is shorter and more curved.

American Avocets feed by wading into shallow pools and skimming food from the top layer of the water by swinging their bills from side to side like a scythe. Often, several will feed abreast, moving through the water step by step and swinging their bills in unison. As they feed, their bills remain slightly open to catch the tiny invertebrates stirred up by the sideways movement. Occasionally, American Avocets swim into deeper water, and even tip up like ducks. They are better swimmers than Black-necked Stilts *(Himantopus mexicanus)* because they have more webbing between their toes.

Most American Avocets breed throughout the interior West at alkali lakes and marshes, then winter along the southern coasts, including that of southern California. They also breed locally in small colonies from Ventura County south to the Mexican border.

SIZE: Length 18 in. **ADULT BREEDING:** Mar.–Aug. Sexes similar. Similar to adult nonbreeding, but apricot cinnamon blush on head and neck. **ADULT NONBREEDING:** Sept.–Feb. Sexes similar. Pale gray head and neck, white underparts. Black-and-white pattern on back. Blue gray legs. Thin, upturned bill (curvature more pronounced in female's bill). **STATUS:** Fairly common migrant and winter visitor from Ventura south, and fairly common to uncommon from Santa Barbara north. Uncommon local summer resident from Ventura County south. **(B)**

SANDPIPERS
(Scolopacidae)

When we think of shorebirds, we tend to picture the sandpipers—the largest family of shorebirds. Many sandpipers are already familiar as the companions of our beach walks. Who hasn't stopped to wonder at their varied bill shapes, or admire the beauty of their fast-flying flocks?

It's difficult to generalize about sandpiper characteristics. Most have slender bills that are either long or short, downcurved, upcurved, or straight, depending upon their feeding habits. Most have relatively long legs in comparison to their bodies, but some have very short legs. Most have slender bodies, but some are chunky. In size, they range from the 26-inch Long-billed Curlew *(Numenius americanus)* to the 6-inch Least Sandpiper *(Calidris minutilla)*.

Most sandpipers hunt for food by feeling for it in the mud. Their bill tips possess sensitive nerve endings that allow them to probe for buried animals they cannot see.

One group of sandpipers, nicknamed the "rockpipers," feeds in a completely different way: they creep about on surf-washed rocks dislodging tiny morsels with their bills. Wandering Tattlers *(Heteroscelus incanus)*, Ruddy and Black Turnstones *(Arenaria interpres* and *A. melanocephala)*, and Surfbirds *(Aphriza virgata)* belong to this group.

Another kind of sandpiper, the phalarope, twirls around in tight circles on the surface of the water, stirring up prey and then stabbing it with a needlelike bill. Wilson's and Red-necked Phalaropes *(Phalaropus tricolor* and *P. lobatus)* are in this group.

Although many species of sandpiper are migrants or winter visitors along southern California's shores, only one, the Spotted Sandpiper *(Actitis macularia)*, remains to breed here.

GREATER YELLOWLEGS	*Tringa melanoleuca*
LESSER YELLOWLEGS	*T. flavipes*

(Pls. 61, 62)

It is fortuitous when a bird's name reflects its appearance; so often it does not. These two shorebirds—the yellowlegs—have yellow legs in all plumages. They forage on the margins of small pools in salt marshes or on the tidal mudflats of estuaries, avoiding the open beach. Yellowlegs do not probe, as most sandpipers, but nab swimming insects, crustaceans, or even small fish by prancing through water, frequently changing direction to stir up prey.

The **GREATER YELLOWLEGS** is a tall sandpiper with bright yellow legs, about the size of a Willet *(Catoptrophorus semipalmatus)*. The bird's bill is long—one-and-a-half times the length of its head—and turns up very slightly. The bill is pale at the base, black at the tip. The call of the Greater Yellowlegs truly sets it apart: a loud and unmistakable cry "tew-tew-tew" that rings out over the marsh.

The **LESSER YELLOWLEGS** is a medium-sized sandpiper with bright yellow legs, approximately the size of a Killdeer *(Char-*

Plate 61. Greater Yellowlegs: adult nonbreeding.

adrius vociferus). Not only is the Lesser Yellowlegs smaller than the Greater, its bill is shorter, equal to the length of the bird's head. The bill is all black and straight. The call of the Lesser Yellowlegs — a soft "tew-tew" — is not loud like that of the Greater Yellowlegs. When either bird is encountered separately, without the other nearby for comparison, identification is sometimes a challenge.

Both species breed in boreal forests and winter from the southern states through South America. At all seasons, Greater Yellowlegs are more abundant than Lesser Yellowlegs along our coast. The best time to see Lesser Yellowlegs is in August and September, when they pass through en route to wintering grounds farther to the south.

GREATER YELLOWLEGS SIZE: Length 14 in. **ADULT BREEDING:** Mar.–Aug. Sexes similar. Similar to adult nonbreeding, but heavily barred underparts. **ADULT NONBREEDING:** Aug.–Mar. Sexes similar. Mottled gray above, white below. Whitish rump and tail. Yellow legs. Long bill, pale at base, turns up slightly. **STATUS:** Fairly common migrant and winter visitor.

Plate 62.
Lesser Yellow-
legs: adult
nonbreeding.

LESSER YELLOWLEGS SIZE: Length 10.5 in. **ADULT BREEDING:** Apr.–
Sept. Sexes similar. Similar to adult nonbreeding, but lightly
barred underparts. **ADULT NONBREEDING:** Oct.–Mar. Sexes similar.
Mottled gray above, white below. Whitish rump and tail. Yellow
legs. Straight, all-dark bill. **STATUS:** Fairly common fall migrant.
Uncommon local winter visitor and spring migrant.

WILLET   *Catoptrophorus semipalmatus*
(Pl. 63)

Among the Marbled Godwits *(Limosa fedoa)* and Black-bellied
Plovers *(Pluvialis squatarola)* along the beach, a tall, plain gray
shorebird with a medium-length straight bill methodically
probes the sand for crabs and small mollusks. Sometimes the bird
hurries down to the surf to follow the Sanderlings *(Calidris alba)*
as they feed in the wake of the restless waves. At other times, this
drab shorebird forages on the upper beach in the dry sand, picking at terrestrial insects and their larvae. During high tide it roosts
in flocks, often standing on one leg with bill tucked.

The Willet is common and widespread as a winterer in southern California. It is a generalist, adapting well to a variety of habitats, whether at rocky shores, sandy beaches, tidal mudflats, or salt
marshes.

Willets spook easily. A birder approaching too closely or a jog-
ger on the run puts them up. As they fly away, the Willets trans-

Plate 63.
Willet: adult
nonbreeding.

form themselves from the most nondescript shorebirds on the beach to the showiest and the loudest. The beautiful black-and-white pattern of the Willets' open wings in flight flashes across the beach. Their screeching "kerrRAK, kerrRAK, kerrRAK" call rouses the other dozing shorebirds, sounding the alarm all along the strand.

SIZE: Length 15 in. **ADULT BREEDING:** Apr.–Aug. Sexes similar. Heavily mottled above, whitish below. **ADULT NONBREEDING:** Sept.–Mar. Sexes similar. Uniform gray above, whitish gray below. Black wings with wide white stripes in flight. Blue gray legs. Straight, dark bill. **STATUS:** Common migrant and winter visitor. Uncommon summer visitor.

WANDERING TATTLER   *Heteroscelus incanus*
(Pl. 64)

During fall and spring migration, the rocky points and headlands along our coast serve as stopping points for shorebirds bound south or north in their wanderings. Of the rockpipers—those sandpipers frequenting rocky habitats—one is taller than the

Plate 64. Wandering Tattler: adult nonbreeding.

others; its nonbreeding plumage is a dull slate gray, which blends neatly with the surrounding mussel-covered rocks. As it walks, this sandpiper often teeters slightly, reminiscent of a Spotted Sandpiper *(Actitis macularia)*.

This is the Wandering Tattler, a solitary species rarely seen in flocks. Tattlers seek breakwaters and jetties on the outer coast for feeding; they peck at small mollusks and marine worms among the algae; or, if the tide is out, they may forage on adjacent wet sand. They are noisy, giving high-pitched calls that carry above the crashing waves.

Wandering Tattlers are world travelers, breeding on mountain streams in southern Alaska and Siberia, then migrating southward to winter along the Pacific coast from southern California to the tip of South America, as well as to New Guinea and Australia.

SIZE: Length 11 in. **ADULT BREEDING:** Apr.–Oct. Sexes similar. Similar to adult nonbreeding, but underparts heavily barred. **ADULT NONBREEDING:** Oct.–Mar. Sexes similar. Entirely slate gray above. Whitish eyebrow. Gray wash on breast, white below. Yellowish legs. **STATUS:** Uncommon to fairly common migrant (especially in spring). Uncommon local winter visitor.

SPOTTED SANDPIPER   *Actitis macularia*

(Pl. 65)

At the edges of lagoons and along the muddy shores of sheltered bays, a small sandpiper investigates the piles of seaweed and debris. Sometimes it hops up and walks along a driftwood log or explores along the rocks. As it searches for flies and beetles, this sandpiper continually teeters the rear end of its body up and down. Aside from the Wandering Tattler *(Heteroscelus incanus)*, whose teetering is not as pronounced, the Spotted Sandpiper is unique among shorebirds in its constant tail-bobbing motion, making it easy to identify. Many Spotted Sandpipers we see are in nonbreeding plumage and do not have spots.

Plate 65. Spotted Sandpiper: adult breeding.

The Spotted Sandpiper feeds singly, not in flocks. It does not probe the mud but darts at prey on the surface. When flushed, this sandpiper flies low over the water with stiff wing beats, its wings held in a shallow, downcurved arc.

Spotted Sandpipers employ a variety of mating systems—some populations being monogamous and others polyandrous (in which the females mate with several males and each male raises a brood). Furthermore, the females are larger, more aggressive, and arrive on the breeding grounds first.

The Spotted Sandpiper has a huge range — breeding from the southern edge of the Arctic to the southern states and wintering all the way to southern South America. It nests in a variety of habitats from sea level to the shores of alpine lakes. In our region, it is a scarce nester along rivers or at freshwater ponds.

SIZE: Length 7.5 in. **ADULT BREEDING:** Apr.–Aug. Sexes similar. Similar to adult nonbreeding, but large black spots on breast and flanks. **ADULT NONBREEDING:** Aug.–Mar. Sexes similar. Brownish gray above, brown wash across breast, white underparts. Whitish eyebrow above black eye-line. Pale pink legs. **STATUS:** Fairly common migrant and winter visitor. Rare local resident. **(B*)**

WHIMBREL   *Numenius phaeopus*
(Pl. 66)

When the tide is running high on the open beach and many of the shorebirds stop feeding and fly to the marsh to roost, a good-sized dusky brown shorebird with a downcurved bill remains on the dry sand. It patrols the upper beach among the piles of wrack. The bird uses its decurved bill to push aside clumps of seaweed and uncover prey.

Closely related to the Long-billed Curlew *(Numenius amer-*

Plate 66. Whimbrel: adult.

icanus), the Whimbrel is distinguished from the Curlew by the alternating dark and light stripes on its crown.

The Whimbrel is found in a variety of coastal habitats, but it typically works the drift line in the drier sand. Unlike some sandpipers, the Whimbrel forages visually, sometimes visiting rocky shores for Striped Shorecrabs *(Pachygrapsus crassipes),* or lawns for insects.

Whimbrels migrate up and down our coast on their way to and from their high Arctic nesting grounds. They spend the winter from San Francisco south as far as Chile. In spring, Whimbrels fly high in V formation flocks, crying out in wild, rolling "turr, turr" notes as they pass overhead.

SIZE: Length 17.5 in. **ADULT:** Sexes similar. All-over mottled grayish brown. Black crown stripes and black eye line. Decurved bill. **STATUS:** Common migrant and uncommon to fairly common local winter visitor. Uncommon summer visitor.

LONG-BILLED CURLEW   *Numenius americanus*
(Pl. 67)

Around the edges of sheltered mudflats and estuaries or in flooded pastures—but seldom on the open beach—a tall sandpiper probes the ground with its exceptionally long, gradually decurved bill, often plunging its bill to the base.

The Long-billed Curlew derives its Latin species name from its crescent-shaped bill: the Greek word *noumenios* means "of the new moon." Its unusual bill seems to have evolved for feeding on its wintering grounds, where it captures Ghost Shrimp *(Callianassa californiensis)* and Blue Mud Shrimp *(Upogebia pugettensis)* deep in their burrows in tidal mudflats along the southern California coast. In one study conducted at Morro Bay, Long-billed Curlews probed deeper in the mud than both Long-billed and Short-billed Dowitchers *(Limnodromus scolopaceus* and *L. griseus)* and Marbled Godwits *(Limosa fedoa).* Interestingly, female Long-billed Curlews have longer bills and are larger than the males.

Long-billed Curlews, unlike most of our wintering shorebirds, nest in the interior short-grass prairies of the western Great Plains and the Great Basin, not in Arctic regions. They are short-distance migrants compared to the Whimbrel *(Numenius phaeopus)* and other sandpipers.

Plate 67. Long-billed Curlew: adult.

SIZE: Length 23 in. **ADULT:** Sexes similar. Cinnamon speckled with brown above, pale buff below. In flight, cinnamon wing linings. Bill long (5 to 8 inches), decurved, basal one-third pale pink. **STATUS:** Uncommon to fairly common migrant and winter visitor.

MARBLED GODWIT   *Limosa fedoa*
(Pl. 68)

This large shorebird with its long, pink-based bill and cinnamon brown plumage resembles a Long-billed Curlew *(Numenius americanus)*, except that its bill turns slightly upward, not downward like the Curlew's. Marbled Godwits are long-legged waders that probe in the wet sand on beaches and tidal mudflats. They gather in medium-sized flocks to feed, then fly to a sandbar or dike to roost, often sleeping while standing on one leg.

Marbled Godwit feeding schedules are hitched to the tides. When the tide is in, they roost. When the tide ebbs, regardless of whether it is day or night, they search the mudflats for polychaete worms, small mollusks, and crabs.

In spring, Marbled Godwits fly to the prairie grasslands of the northern United States and southern Canada to nest. Most of the

Plate 68. Marbled Godwit: adult nonbreeding.

year, however, they can be seen on southern California and Mexican beaches, reaping the bountiful harvest hidden in the mud.

SIZE: Length 18 in. **ADULT BREEDING:** Apr.–Sept. Sexes similar. Similar to nonbreeding but extensive dark barring above and below. **ADULT NONBREEDING:** Oct.–Mar. Sexes similar. Mottled dark and light upperparts on a background of cinnamon brown. Buff underparts, cinnamon wing linings. Long bill (pink on the basal half, black on the terminal half) turns up slightly. **STATUS:** Common migrant and winter visitor. Uncommon summer visitor.

RUDDY TURNSTONE   *Arenaria interpres*
BLACK TURNSTONE   *A. melanocephala*
(Pls. 69, 70)

Turnstones are named for their feeding behavior: they use their short, pointed bills to flip over stones, seaweed, pebbles, and shells while searching for prey. Their bills are slightly upturned, perfect for poking into the crevices of rocky shores.

The **RUDDY TURNSTONE** is a short-legged, squat bird with round, brownish patches on its breast, white underparts, and orange legs. Ruddy Turnstones are found on rocky shores and pebbly flats where they seek insects and larvae. Often, the Ruddy Turnstone inspects the seaweed mounds or scurries across the dry

Plate 69. Ruddy Turnstone: adult nonbreeding.

sand pushing its bill in front of it like a little shovel to uncover invertebrates lurking beneath the surface. The Ruddy is less tied than the Black Turnstone to the exclusive use of rocky shores.

Although Ruddy Turnstones are not as frequently seen along southern California's shores as Black Turnstones, they are found worldwide, being Arctic breeders in both hemispheres, and winterers on temperate coasts. Black Turnstones, in contrast, have a purely western distribution: they breed in western Alaska and winter only along the Pacific coast.

The **BLACK TURNSTONE** is at home in the splash zone, where the surf spray wets its feathers and the rocks are steep. Black Turnstones feed mostly on Acorn Barnacles *(Balanus glandula)* by inserting their closed bill into the shell, opening and prying them apart. Black Turnstones are short-legged birds with blackish upperparts and breasts and white bellies. Well camouflaged, they creep about on the dark rocks of breakwaters or sea stacks near the coast. If an intruder startles the birds, they flush at once, becoming a striking assemblage of flashing black-and-white wings. Researchers hypothesize that the beautiful black-and-white pattern of the Black Turnstones' back and wings in flight may be an adaptation to help them escape predators. By suddenly overwhelming a predator with a fluttering group of black-and-white shapes, the Black Turnstones can disrupt dangerous situations.

Plate 70. Black Turnstone: adult nonbreeding.

RUDDY TURNSTONE SIZE: Length 9.5 in. **ADULT BREEDING** Apr.–Sept. Sexes similar. Black-and-white head, black round patches on breast, rusty back. Orange legs. **ADULT NONBREEDING:** Sept.– Apr. Sexes similar. Round, brownish patches on breast, brown upperparts, white below. Orange legs. **STATUS:** Fairly common migrant. Uncommon local winter visitor.

BLACK TURNSTONE SIZE: Length 9.25 in. **ADULT BREEDING** Apr.–Aug. Sexes similar. Similar to adult nonbreeding, but white spot behind bill, white eyebrow, white marks on breast. **ADULT NONBREEDING:** Aug.– Apr. Sexes similar. Black upperparts and breast, white underparts. Dark legs. **STATUS:** Fairly common migrant and winter visitor.

SURFBIRD   *Aphriza virgata*
(Pl. 71)

From the cliffs at La Jolla to the jetties at Playa del Rey to the surf-splashed ledges of Montana de Oro, birders scan the rocky outcrops looking for a gray sandpiper with a short bill that pauses along the coast in migration and stays sparingly in winter.

They seek the Surfbird—another of the rocky shorebird specialists—frequently found in the company of Black Turnstones (*Arenaria melanocephala*). The Surfbird thrives in the splash zone, as it picks off mussels, limpets, and barnacles from the rocks with a sideways pull of its short bill.

Surfbirds have a remarkably long, narrow wintering range: they are found only on the Pacific coast from southern Alaska all the way to the Strait of Magellan, a distance of some 10,870 miles. In spring, they make their way northward to nest beside rushing rivers on slopes above timberline in Alaska. So remote are Surfbird nests, they were not discovered until 1926, when one was finally located in what is now Denali National Park in Alaska.

Surfbirds are most easily seen along our seacoasts during spring migration, when they stop over singly or in small flocks on wave-beaten rocks. At this season, they often show bright, coppery-colored back (scapular) feathers as they begin to acquire nuptial plumage.

Plate 71. Surfbird: adult breeding.

SIZE: Length 10 in. **ADULT BREEDING:** Mar.–Aug. Sexes similar. Similar to adult nonbreeding, but copper-colored back (scapular) feathers and more dark spots. **ADULT NONBREEDING:** July–Apr. Sexes similar. Slate gray above and across breast, white below with faint spots. Yellow legs. Yellow at base of bill. **STATUS:** Fairly common spring migrant. Uncommon local fall migrant and winter visitor.

RED KNOT   *Calidris canutus*

(Pl. 72)

The Latin species epithet for this sandpiper, *canutus,* commemorates the Danish King Canute, or Knut, so the legend goes. Having dined on a succulent dish of the bird, the king showed a marked interest in it ever after.

Plate 72. Red Knot: adult nonbreeding.

Red Knots are like giant, gray Sanderlings *(Calidris alba):* stocky with short bills in proportion to their bodies. They differ from Sanderlings, however, in that they do not feed by running up and back with the waves. Red Knots probe for mollusks where extensive tidal mudflats lie protected from the waves at estuaries and lagoons.

Red Knots are an uncommon sight anywhere along the southern California coast, but they do overwinter from Orange County southward, where they can be found among the other shorebirds that probe sheltered mudflats.

SIZE: Length 10.5 in. **ADULT BREEDING:** May–Aug. Sexes similar. Brick red sides of face, breast, and belly. Dark brown upperparts with reddish feather edges. **ADULT NONBREEDING:** Sept.–Apr. Sexes similar. Medium gray upperparts, white below. Black legs. Short, black bill. **STATUS:** Rare to uncommon migrant. Fairly common winter visitor from Orange County south, rare elsewhere.

SANDERLING   *Calidris alba*

(Pl. 73)

Along the sandy beach, a flock of small, whitish sandpipers runs swiftly down to the edge of the surf. In the wake of the retreating wave, each bird probes frantically at the wet sand. Then, as the next wave surges up, the sandpipers run rapidly in front of it. With their stubby, fast-moving legs beneath plump bodies, they look like mechanical toys. These are the Sanderlings, the most visible sandpipers of southern California's beaches. Even a casual observer notices the Sanderlings.

Plate 73. Sanderling: adult nonbreeding.

A Sanderling spends most of its life at the shore, following the fringes of the waves in endless rhythm up and back. It feeds and roosts in flocks, watchful for marauding falcons from above and the next morsel of food hidden beneath the sand below. When danger threatens, it flees with other Sanderlings in whirling flight down the beach, only to alight again and resume the frantic searching of the wet sand. At high tide, Sanderlings retreat to wet rocks, where they are surprisingly agile at gleaning crustaceans from the algae.

In many ways, the Sanderling is the universal shorebird: a high Arctic breeder around the world that winters on sea beaches in both hemispheres and on all the continents. It has one of the widest latitudinal wintering ranges of any shorebird in North

America—except perhaps the Surfbird *(Aphriza virgata)*—found from 50 degrees north latitude (British Columbia) to 50 degrees south latitude (southern Chile and Argentina).

Banding studies show that some Sanderlings that end up wintering along the coast in Peru and Chile have migrated south along the Atlantic coast or through coastal Texas in fall to reach the southern Pacific coast of South America. In spring, though, they fly north along the U.S. Pacific coast on their return journey to the Arctic—an incredibly long, oval-shaped migration route.

SIZE: Length 8 in. **ADULT BREEDING:** May–July. Sexes similar. Reddish brown wash on head and neck, darker back. **ADULT NONBREEDING:** Aug.–Apr. Sexes similar. Upperparts pale grayish white, dusky wedge at bend of wing. White underparts. **STATUS:** Common migrant and winter visitor.

WESTERN SANDPIPER   *Calidris mauri*
LEAST SANDPIPER   *C. minutilla*
(Pls. 74, 75)

On a warm August afternoon when the tide is ebbing and the mudflats at shallow bays are exposed, thousands of tiny sandpipers spread themselves over the estuary. Some are belly deep in puddles, others scurry about on the drier sand.

Plate 74. Western Sandpiper: adult breeding.

These are the **WESTERN SANDPIPER** and the **LEAST SANDPIPER**, the two smallest sandpipers found along our coast and the littlest of the genus *Calidris,* collectively called "peeps" by birders. They dab and probe for the plentiful food in the black ooze. Working with their bills as fast as they can, the peeps refuel for the next leg of their migratory journey. Separating the seething mass of sandpipers into species is a daunting task, but with time and patience the peeps can be distinguished from one another.

Plate 75. Least Sandpiper: adult breeding, worn.

The best way to study Western and Least Sandpipers is with the help of a spotting scope. Through the scope, you can see that the Western Sandpiper is a bit taller than the Least. Its bill is thicker at the base and tapers to a point, and it has black legs.

The Least Sandpiper has even shorter legs than the Western, giving it a "hunkered down" appearance as it runs about on the sand. The bill of the Least Sandpiper is thin and needlelike, not tapered like the Western's. The Least Sandpiper's legs are a yellowish green color, lighter than the black legs of the Western.

Both Least and Western Sandpipers are common migrants along our coast in spring and fall. In winter, Least Sandpipers are widespread throughout our region, but Western Sandpipers are usually found only at the larger tidal bays such as South San Diego Bay, Upper Newport Bay, and Morro Bay (see also page 41).

WESTERN SANDPIPER SIZE: Length 6.5 in. **ADULT BREEDING:** Mar.–Aug. Sexes similar. Rufous feathering on crown, ear patch (auricular), and upper back (scapular) feathers. Extensive dark spotting across breast. **ADULT NONBREEDING:** Aug.–Mar. Sexes similar. Pale gray above, whitish underparts. Black legs. **STATUS:** Common migrant. Fairly common local winter visitor.

LEAST SANDPIPER SIZE: Length 6 in. **ADULT BREEDING:** Apr.–Aug. Sexes similar. Similar to adult nonbreeding, but richer brown on upperparts. **ADULT NONBREEDING:** Aug.–Mar. Sexes similar. Brownish gray above with brownish wash over breast. Yellowish green legs. **STATUS:** Common migrant and winter visitor.

DUNLIN   *Calidris alpina*
(Pl. 76)

When fall days grow shorter in September and many other shorebird species depart for destinations farther south, a new peep (see page 159) shows up. This sandpiper is larger than the Western

Plate 76. Dunlin: adult nonbreeding.

and Least Sandpipers *(Calidris mauri* and *C. minutilla)* and has a longer bill, which droops slightly at the end. This is the Dunlin, another long-distance migrant just arrived from northern Alaska.

In fall and winter plumage, Dunlin are plain, dull gray with grayish breasts, but when a flock of Dunlin moves through in spring migration each bird sports a striking black patch in the middle of its white belly and lots of rich rufous feathering on the back.

Along with Western Sandpipers, Dunlin are the most numerous shorebird species to use the southern California coast as a stopover in fall. Unlike most Western Sandpipers, however, Dunlin are common winterers here, too.

Wintering Dunlin flocks twist and bank in the air with superb coordination. When flushed, they whirl in one direction, then abruptly change to another. Even when Peregrine Falcons *(Falco peregrinus)* and Merlins *(F. columbarius)*—their chief predators—are not threatening, Dunlin fly in tight formation over the mudflats.

SIZE: Length 8.5 in. **ADULT BREEDING:** Apr.–Aug. Sexes similar. Extensive rufous feathering on back, distinct black patch on white belly. **ADULT NONBREEDING:** Sept.–Mar. Sexes similar. Grayish upperparts, grayish upper breast, whitish underparts. Black legs. Drooped, medium-length bill. **STATUS:** Common to fairly common migrant and winter visitor.

SHORT-BILLED DOWITCHER   *Limnodromus griseus*
LONG-BILLED DOWITCHER   *L. scolopaceus*
(Pls. 77, 78)

By studying the feeding shorebirds at tidal mudflats anywhere along the coast, a birder sorts through the sandpipers: the smallest are peeps, the tallest are Willets *(Catoptrophorus semipalmatus),* Marbled Godwits *(Limosa fedoa),* and Long-billed Curlews *(Numenius americanus).* But what about those medium-sized shorebirds in that cluster over there? They are chunky sandpipers with long, straight bills, which they jab continuously into the mud. This "sewing machine movement"—feeding by pumping their bills rapidly up and down—is characteristic dowitcher behavior.

Dowitchers are gregarious. They fly in wheeling flocks, suddenly dropping down to land en masse. Like the snipe, they have oversized bills in proportion to their round, stocky bodies.

Telling **LONG-BILLED DOWITCHERS** from **SHORT-BILLED DOWITCHERS** is very tricky, especially in nonbreeding plumage. Up until 1950, they were mislabeled even in museums. Their bills overlap in dimensions, so bill length is not a reliable field mark.

The best way to distinguish them is to learn to recognize their calls. The Long-billed has a high, single "keek" call. (Caution:

Plate 77. Short-billed Dowitcher: juvenile.

Plate 78. Long-billed Dowitcher: adult nonbreeding.

These "keek" calls can be strung together in a series when the bird is flushed.) The Short-billed Dowitcher gives a low, mellow "tu-tu-tu" call.

In winter on southern California's coast, Short-billed Dowitchers are scarce, although they may be found in some tidal salt-water environs in Ventura, Orange, and San Diego Counties.

Long-billed Dowitchers, on the other hand, are common most everywhere in both fresh- and saltwater habitats in winter.

SHORT-BILLED DOWITCHER SIZE: Length 11 in. **JUVENILE:** Sexes similar. Buff-colored bars on dark tertial feathers. **ADULT BREEDING:** Apr.–July. Sexes similar. Cinnamon red foreneck, spots on sides of breast, whitish belly. **ADULT NONBREEDING:** Aug.–Apr. Sexes similar. Pale gray brown above with pale line above eye. Gray throat and breast, whitish below. Yellowish green legs. **STATUS:** Common migrant. Uncommon to fairly common local winter visitor.

LONG-BILLED DOWITCHER SIZE: Length 11.5 in. **JUVENILE:** Sexes similar. Lacks buff-colored bars on dark tertial feathers. **ADULT BREEDING:** Apr.–Aug. Sexes similar. Similar to Short-billed Dowitcher, but richer colored red underparts (including belly), dark barring on sides of breast. **ADULT NONBREEDING:** Aug.–Apr. Sexes similar. Similar to Short-billed Dowitcher, but slightly darker on back and breast. **STATUS:** Common migrant and winter visitor.

WILSON'S SNIPE   *Gallinago delicata*
(Pl. 79)

The startling explosion of a shorebird flushed from underfoot in a rain-soaked field breaks the morning's silence. The bird leaves

Plate 79. Wilson's Snipe: adult.

the ground with swift wings beating the air. It zigzags in escape flight and finally climbs high into the sky and disappears. That fleeting glimpse is likely to be a birder's brief introduction to the Wilson's Snipe.

Formerly called Common Snipe, the Wilson's Snipe is a bird of freshwater marshes, ditches, and flooded fields, where its buff-colored plumage matches the weeds and grasses. It crouches motionless in perfect camouflage, or probes the ground with its extremely long bill. The snipe resembles a dowitcher in its body proportions, with a chunky body and an oversized bill.

Solitary in its habits, the Wilson's Snipe may gather with others of its kind when feeding is particularly good, such as after a heavy rain.

SIZE: Length 10.5 in. **ADULT:** Sexes similar. Striped tan-and-black head. Several prominent buff-colored stripes on black speckled back. Extremely long, straight bill. **STATUS:** Uncommon to fairly common migrant and winter visitor.

WILSON'S PHALAROPE
RED-NECKED PHALAROPE

Phalaropus tricolor
P. lobatus

(Pls. 80, 81)

In deeper pools on estuarine mudflats and in shallow lagoons, these little shorebirds pirouette around and around on the surface of the water, their bills dabbing at the larvae stirred up by the swimming movements of their lobed feet. Their singular feeding habits and dainty mien proclaim the phalaropes' presence among a gathering of other sandpipers.

Phalaropes' breeding habits embody the reversal of the sex roles: the females are more brightly colored, and they conduct most of the courtship cycle from start to finish, pausing only to lay the eggs. The males are entirely responsible for incubation and brood rearing. Also, phalaropes are polyandrous, a mating system in which each breeding season, females mate with more than one male and lay multiple broods. Each male incubates one batch of eggs and raises one brood.

WILSON'S PHALAROPES use the whirligig method of feeding on water, but they are equally at home on land. This phalarope walks quickly along muddy flats, stooping and darting at prey, rather like a yellowlegs. Female Wilson's Phalaropes are one of the earliest shorebirds to return in fall migration, often first appearing on our shores by mid-June. They have a long journey ahead.

Plate 80. Wilson's Phalarope: adult male breeding, worn.

Wilson's Phalaropes undergo an interesting migration in fall. They breed at freshwater ponds in western North America (not in Arctic regions). Many, thought not all, gather at large saline waters, such as Mono Lake, Great Salt Lake, and San Francisco Bay. From these staging areas, most Wilson's Phalaropes make long-distance flights to spend the winter at saline lakes in the highlands of Bolivia, Chile, and Argentina. Since there are no records of this species in Central and northern South America during fall, ornithologists believe Wilson's Phalaropes make these migration flights — often totaling 54 hours — without stopping at all en route.

RED-NECKED PHALAROPES, in contrast, are much less land-bound than Wilson's. In late summer, they appear along our coast spinning on the surface of wastewater treatment ponds or brackish lagoons, seldom walking on mudflats. Note that whatever the season, this bird's back has a patterned appearance of light streaks on dark, in contrast to the rather uniform gray of the back of the Wilson's Phalarope.

When Red-necked Phalaropes migrate south from subarctic nesting areas, they journey to the seas of the Southern Hemisphere to the Humboldt Current off Peru and Ecuador. For the next nine months, these diminutive birds float in rafts of thousands on the open ocean at locations on the edges of tidal rips, where plankton is plentiful. They are visual foragers, stabbing and picking prey from the water's surface as they create their own miniature upwellings by using their feet to propel them round and round.

Plate 81. Red-necked Phalarope: adult female breeding.

In April, Red-necked Phalaropes occasionally pass close enough to southern California shores on their northward migration flights to produce incredible sightings from local promontories of thousands of birds going by in a matter of hours.

WILSON'S PHALAROPE SIZE: Length 9.25 in. **ADULT MALE BREEDING:** Apr.–July. Duller than female, lacks dark stripe down neck. **ADULT FEMALE BREEDING:** Apr.–July. Pale gray crown and nape. Dark stripe through eye and down neck, chestnut wash on sides of neck. Black legs. **ADULT NONBREEDING:** Aug.–Mar. Sexes similar. Pale gray upperparts, white face, white underparts. Pale yellow legs. Long, thin dark bill. **STATUS:** Fairly common to common fall migrant. Uncommon to fairly common spring migrant.

RED-NECKED PHALAROPE SIZE: Length 7.75 in. **ADULT MALE BREEDING:** Apr.–July. Duller than female, lacks dark face. **ADULT FEMALE BREEDING:** Apr.–July. Dark gray upperparts streaked with buff on back. White throat, chestnut patches on sides of neck. **ADULT NONBREEDING:** Aug.–Apr. Sexes similar. Medium gray upperparts with pale streaks, white underparts. Dark mark through eye. Thin bill shorter than Wilson's Phalarope. **STATUS:** Common fall migrant. Fairly common spring migrant, but numbers vary from year to year.

GULLS, TERNS, and SKIMMERS (Laridae)

Please see "Gull Identification," at the back of the book, for further information on the study of gulls.

Gulls, terns, and skimmers belong to a large family of seabirds called the Laridae, which have the three front toes webbed. Most of the larids live close to water, but each group has distinctive feeding behaviors.

Gulls symbolize the seashore. No other group embodies the spirit of the sea like gulls—their wild cries evoking a life of adventure on the high seas that we may never experience.

Yet, gulls live comfortably side by side with humans, often near the coast but sometimes far inland. Gulls stand on our beaches, perch on light poles along coast roads, and sidle around fishing docks looking for handouts. Gulls wait at playgrounds for leftover lunch scraps, follow the plow to feed on insects, and scavenge garbage at dumps.

A majority of gulls are omnivorous, eating anything from garbage to carrion to the young or eggs of other seabirds. At sea, most do not dive for prey but circle high in the sky, then land and snatch fish from the surface of the water. They have learned to drop clams and other shellfish from a height onto rocks along the shore, then descend to eat the meat from the broken shells.

Terns look like gulls from afar because they are white, gray, and black and are found near water. Studied closely, terns are quite different. Terns don't soar the way gulls do; with each downstroke of their pointed wing tips, they appear to bounce slightly upward as they sail gracefully along.

Terns do not scavenge. When hunting, they cruise in flight with bill directed downward (not straight out like the gulls'), on the prowl for a meal. Sighting a live fish, squid, or shrimp, they plunge headlong into the water from above.

Terns have short necks and stubby legs. In a group of mixed terns and gulls resting on the beach, the terns' squat silhouettes set them apart from the taller gulls. So short are the tern's legs, they scarcely raise the bird an inch above the sand.

The skimmer, a striking member of the Laridae, is famed for

its oversized bill, which is used to skim the surface of calm waters in scooping up small prey.

Southern California's coast is an excellent place to observe eight species of gull, five species of tern, and the Black Skimmer *(Rynchops niger)*. Fall and winter are the best times to study gulls; terns and skimmers are most numerous in summer.

BONAPARTE'S GULL   *Larus philadelphia*
(Pl. 82)

A small gull with a pale gray back and a white head with a dark spot behind the eye swims on sheltered lagoons or at wastewater treatment plants near the coast. As it sits on the water, it jabs at invertebrates floating near the surface, in the manner of a phalarope. When it takes to the air, its pointed wings and agile flight resemble those of a tern more than a gull.

Plate 82. Bonaparte's Gull: adult nonbreeding.

The Bonaparte's Gull is the smallest and most delicate of our gulls. It was named after a famous French zoologist, Charles Lucien Bonaparte—the younger brother of Napoleon—who spent time in Philadelphia and was an important figure in early North American ornithology.

Bonaparte's Gulls differ from other common southern California gulls in appearance and behavior. In breeding plumage, they acquire dark hoods, unlike the white heads of most other species. Also, Bonaparte's Gulls are not found at garbage dumps, seldom soar on set wings, and eschew urban habitats. Instead, they make graceful aerial dips at small prey found in tidal inlets, beach outfalls, or kelp beds.

In spring, we see Bonaparte's Gulls as they migrate north along the Pacific coast. They are en route to breeding grounds in the boreal forests of Canada and Alaska, where, interestingly, they choose to nest in trees.

SIZE: Length 13.5 in. **ADULT BREEDING:** Apr.–Aug. Sexes similar. Similar to nonbreeding, but whole head black. **ADULT NONBREEDING:** Aug.–Apr. Sexes similar. White head, dark spot behind eye, pale gray mantle, black edging on outer primaries. Pink legs. Black bill. **STATUS:** Fairly common migrant and winter visitor.

HEERMANN'S GULL   *Larus heermanni*
(Pl. 83)

A novice who embarks on the study of gulls is grateful for the blood red bill of the Heermann's Gull. No other common gull possesses a red bill, making this species one of the easiest to identify of our gulls. Named after Dr. Adolphus Heermann, a surgeon on one of the U.S. survey teams of the West in the 1850s, the Heermann's Gull is the darkest of the common gulls.

Heermann's Gulls breed in large colonies on islands in the Gulf of California, in Mexico. The largest is at Isla Rasa (since the 1960s, a seabird sanctuary established by the Mexican government), where 90 to 95 percent of the world population of Heermann's Gulls nests — 300,000 individuals in the 1990s. In early summer, when they have finished breeding, the gulls move northward along the Pacific coast, a migration that coincides with that of the Brown Pelican *(Pelecanus occidentalis),* also dispersing northward after nesting.

Heermann's Gulls kleptoparasitize Brown Pelicans by snatching fish from their pouches (see Brown Pelican account). In one study, scientists observed that Heermann's Gulls were more likely to attack adult Brown Pelicans, rather than immatures, after their dives, because the gulls had learned that the adult pelicans are more successful at catching fish.

Heermann's Gulls stick close to the coast, rarely venturing inland. They are often the most numerous gull at the beach in summer.

Plate 83. Heermann's Gull: adult breeding.

SIZE: Length 19 in. **ADULT BREEDING:** Dec.–Aug. Sexes similar. Similar to adult nonbreeding, but with white head. **ADULT NONBREEDING:** Aug.–Feb. Sexes similar. Uniform dark gray body and head. Black legs. Red bill with black tip. **STATUS:** Common year-round visitor except March through May.

MEW GULL · *Larus canus*
(Pl. 84)

Most of the adult gulls that spend the winter on southern California's seacoast have white heads and underparts, gray backs, and black-and-white wing tips. A beginner, therefore, must use other clues to identify each of the gull species within this very general description. Of this white and gray and black group of gulls, then, only one, the Mew Gull, has a short, unmarked yellowish bill in adult plumage. Its dark eye looks large in the rounded head, giving the bird a gentle expression.

Mew Gulls congregate in small flocks at the beach, at flooded pastures, or at wastewater treatment plants. When hunting for food, Mew Gulls hover with legs dangling above the water, similar to Bonaparte's Gulls *(Larus philadelphia)*.

In North America, the Mew Gull breeds only in far northwestern portions of Canada and Alaska; however, in Europe and Asia

Plate 84. Mew Gull: adult nonbreeding.

—where it's called the Common Gull—it breeds throughout northern latitudes. Mew Gulls are one of the latest gull species to arrive on our coast to spend the winter, usually in late October or early November, and most of them have departed by late March.

SIZE: Length 16 in. **ADULT NONBREEDING:** Sept.–Apr. Sexes similar. Smudgy streaking on white head and neck, white body, medium gray back (darker than Ring-billed Gull). Black-and-white wing tips. Yellowish legs. Yellow bill (sometimes dusky at tip). **STATUS:** Fairly common migrant and winter visitor from Ventura County north, uncommon farther south.

RING-BILLED GULL   *Larus delawarensis*
(Pl. 85)

In winter on sandy beaches or on estuarine mudflats, an enormous flock of gulls will sometimes gather to rest. The majority of them will probably be Western or California Gulls *(Larus occidentalis* or *L. californicus),* but after examination through a scope, it is not difficult to pick out several individuals that have a pale gray back and a yellow bill crossed near the tip by a black band. These are adult Ring-billed Gulls, medium-sized gulls smaller than California Gulls and with slightly paler gray backs.

Plate 85. Ring-billed Gull: adult nonbreeding.

From the 1850s to the 1920s, Ring-billed Gulls were in serious trouble—decimated by the millinery trade and persecuted by egg hunters. They have since recovered, and their western population has greatly expanded because of human settlement of the United States and Canada. The spread of agriculture and the creation of garbage dumps has played a part in the growth of Ring-billed Gull numbers.

Ring-billed Gulls are inland nesters. They favor islands in lakes and reservoirs. In a survey from 1994 to 1997, 9,611 to 12,660 pairs of Ring-billed Gulls nested in California, mostly in northeastern California at the Butte Valley Wildlife Area and the Clear Lake and Honey Lake National Wildlife Refuges. After nesting, they migrate to coastal California to winter, where they are commonly spotted in parking lots, playing fields, and landfills, as well as at the beach.

SIZE: Length 17.5 in. **ADULT BREEDING:** Apr.–Sept. Sexes similar. Similar to adult nonbreeding, but lacks head streaks. Red mouth lining (gape). **ADULT NONBREEDING:** Sept.–Apr. Sexes similar. Some gray streaking on white head, pale eye, white underparts, pale gray back. Black-and-white wing tips. Yellow legs. Yellow bill crossed by black ring near tip. **STATUS:** Common migrant and winter visitor. Fairly common summer visitor.

CALIFORNIA GULL   *Larus californicus*

(Pl. 86)

Let's return to that big flock of roosting gulls on the beach in winter. Scanning this daunting collection of white-headed, gray-backed birds, you soon realize that it will be necessary to focus on relative size and bill color in order to differentiate one species from another.

This is not as hard as it seems. The majority of the birds in the flock are larger than a Ring-billed Gull *(Larus delawarensis)* and smaller than a Western Gull *(Larus occidentalis)*. These are Cali-

Plate 86. California Gull: adult breeding.

fornia Gulls, arguably the most common gull along the southern California coast in winter. The key field mark to watch for on adult California Gulls is the yellow bill with a red and black spot near the tip.

California Gulls gained fame in the nineteenth century when they saved the crops of Mormon settlers by devouring invading crickets and grasshoppers. They are the state bird of Utah, commemorated by a statue in Salt Lake City.

California Gulls nest in large colonies on alkali lakes of the interior West, such as Great Salt Lake and Mono Lake. In the same California survey mentioned above, from 1994 to 1997 there were approximately 33,125 to 39,678 pairs of California Gulls nesting in the state, 70 to 80 percent of which nested at Mono Lake. In addition, since 1980, California Gulls have colonized south San Francisco Bay, where they nest on dikes.

California Gulls winter along the Pacific coast, where they sometimes follow fishing boats offshore. Like Ring-billed Gulls, they also visit shopping centers, school playgrounds, park lawns, and municipal landfills.

SIZE: Length 21 in. **ADULT BREEDING:** Mar.–Sept. Sexes similar. Similar to adult nonbreeding, but lacks head flecking. Red mouth lining (gape). **ADULT NONBREEDING:** Oct.–Apr. Sexes similar. White head with brownish flecking, white underparts, medium gray back. Black-and-white wing tips, yellowish green legs. Yellow bill with black band across tip of both mandibles, red dot on lower mandible. **STATUS:** Common migrant and winter visitor. Fairly common summer visitor.

HERRING GULL   *Larus argentatus*
(Pl. 87)

As soon as a birder begins to discriminate one species of adult gull from another among the more common varieties at the beach, along comes a bird that doesn't quite fit. It is a large gull with a pale gray back (similar to the much smaller Ring-billed Gull *[Larus delawarensis]*), but it has a yellow bill with a red spot toward the tip, a pale eye, and pink legs.

You might be tempted to call the bird a Western Gull *(Larus occidentalis),* because it has pink legs and a yellow bill with a red spot. But typical Western Gulls have dark backs; besides, this bird's bill is slender, not thick like that of a Western Gull.

The mystery bird is the Herring Gull, an uncommon southern California gull; however, elsewhere in North America and throughout Europe and Asia, it is widespread and abundant. Many of the Herring Gulls we see are immature nonbreeders, because the adults stay near their breeding grounds farther north throughout the year.

Herring Gulls are opportunistic feeders. Whether hovering over a school of fish offshore, pursuing a fishing boat for leftovers, or scavenging at a landfill, Herring Gulls are aggressive and omnivorous. After many years of coexisting with humans, Herring Gulls grab what they can, prospering by adapting to a variety of habitats.

Plate 87. Herring Gull: adult nonbreeding.

SIZE: Length 25 in. **ADULT BREEDING:** Apr.–Sept. Sexes similar. Similar to adult nonbreeding. Yellow mouth lining (gape). **ADULT NON-BREEDING:** Sept.–Apr. Sexes similar. White head with some flecking, pale eye, white underparts, pale gray back. Black-and-white wing tips. Pink legs. Yellow bill with red spot at tip of lower mandible. **STATUS:** Uncommon migrant and winter visitor.

WESTERN GULL   *Larus occidentalis*
(Pl. 88)

At almost any season along the southern California coast, a large gull with a dark gray back and a thick yellow bill with a red spot near the tip masses in flocks on the sandy beach or perches on light poles or wharf pilings.

A big, bulky bird, the Western Gull is the only gull to breed locally. In spring, Western Gulls crowd the cliffs on the Channel Islands and carefully defend their nests. They also nest on isolated rocks and ledges along the mainland coast, chiefly north of Point Conception.

The Western Gull is a bold, skillful hunter and scavenger. Everything from live fish to dead clams to the eggs and young of other seabirds is fair game. For the most part, Western Gulls are coastal dwellers, content to ride the wind currents above the

Plate 88. Western Gull: adult nonbreeding.

breakers or pause, wings outstretched, buffeted by the onshore breeze. Recently, however, Western Gulls in our region have begun to leave their coastal haunts in favor of exploiting garbage facilities as much as 10 miles inland.

Southern California's Western Gull population increased from 5,500 pairs in the late 1970s to 14,000 pairs in the 1990s. Compared to other North American gull species, however, the Western Gull is rather restricted in its range, breeding only along the Pacific coast of North America from Washington to central Baja California.

In the 1970s, Western Gull colonies in the Southern California Bight attracted public attention when ornithologists observed female-female pairs, which they attributed to a skewed sex ratio brought on by the feminization of male embryos by DDT residues. These chemicals act as estrogen mimics. When chemical waste was banned from ocean outfalls, sex ratios returned to normal.

SIZE: Length 25 in. **ADULT BREEDING:** Feb.–Sept. Sexes similar. Similar to adult nonbreeding. Pink mouth lining (gape). **ADULT NON-BREEDING:** Sept.–Mar. Sexes similar. White head and underparts, dusky eye, dark gray mantle. Pink legs. Thick yellow bill with red spot at tip of lower mandible. **STATUS:** Common resident. **(B)**

GLAUCOUS-WINGED GULL   *Larus glaucescens*

(Pl. 89)

This large gull, one of the "white-headed, gray-backed" group of gulls that appear so similar to the untutored eye, is easily told at all stages by the absence of black in the wing tips.

Plate 89. Glaucous-winged Gull: adult nonbreeding.

The Glaucous-winged Gull is a common winterer in northern California but is much less frequently encountered in southern California. Usually, it is the immatures that venture this far south: they lack any darker coloration in their wing tips, so that their whole body is a creamy beige. In adult plumage, the word "glaucous" perfectly describes the uniform pale grayish white coloration of this gull's back and wing tips.

Glaucous-winged Gulls breed along the coast from Alaska to northern Oregon. In the southern one-third of their range, they sometimes hybridize with Western Gulls *(Larus occidentalis)*, producing intergrades that are difficult to identify.

SIZE: Length 26 in. **ADULT BREEDING:** Feb.–Sept. Sexes similar. Similar to adult nonbreeding, but lacks head streaking. Pink mouth lining (gape). **ADULT NONBREEDING:** Sept.–Mar. Sexes similar. White head flecked with gray, white underparts, pale gray mantle.

Pale gray wing tips. Pink legs. Thick yellow bill with red spot at tip of lower mandible. **STATUS:** Uncommon to fairly common migrant and winter visitor, especially north of Point Conception. Rare in summer.

CASPIAN TERN   *Sterna caspia*
(Pl. 90)

Plate 90. Caspian Tern: adult breeding.

Standing on the boardwalk at Bolsa Chica, a birder sees overhead a large tern with a black cap and a massive, bloodred bill flying toward the nesting islands. The bird struggles to control the wiggling Topsmelt *(Atherinops affinis)* in its bill destined for the youngster back at the nest colony. As the tern passes above, the dark undersides of the wings are clearly visible with each wing beat, and the rasping snarl of its loud "ka-kaow" call, clinches the bird's identification as a Caspian Tern.

Caspian Terns, named for the Caspian Sea where the species was first collected in 1770, are found throughout most of the world. They are the largest and strongest of our terns, resembling gulls with their broad wings and soaring flight.

Caspian Terns are expanding their range throughout North America. Along the southern California coast, they occupy several of the tern nesting colonies, together with Royal and Elegant

Terns *(Sterna maxima* and *S. elegans)* and Black Skimmers *(Ryn-chops niger).* Caspian Terns winter from Santa Barbara County south to Guatemala.

SIZE: Length 21 in. **ADULT BREEDING:** Feb.–Oct. Sexes similar. White head and underparts, pearl gray upperparts, black cap. Heavy bloodred bill with dusky tip. **ADULT NONBREEDING:** Oct.–Feb. Sexes similar. Similar to adult breeding, but black cap paler, flecked with gray. **STATUS:** Fairly common migrant. Uncommon local winter visitor. Fairly common local summer resident from Los Angeles County south. **(B)**

ROYAL TERN   *Sterna maxima*
(Pl. 91)

Beginning in late October and early November on the wide beaches emptied of vacationers and beachcombers, an assemblage of medium-sized terns rests with the gulls on the dry sand. The terns are in their winter, or nonbreeding, plumage and have short black crests, white foreheads, and heavy carrot-colored bills.

The Royal Tern is the most frequently seen medium-sized tern in winter along the coast from Morro Bay south. It is often confused with the very similar Elegant Tern *(Sterna elegans)*—

Plate 91. Royal Tern: adult nonbreeding.

another crested tern with an orange bill—but the Elegant Tern is largely a summer and early fall visitor. The Royal Tern is slightly larger and bulkier than the Elegant, and has a thick, straight bill. The Elegant Tern is more slender bodied; its bill is thinner and appears slightly downcurved.

One of the largest breeding colonies of Royal Terns on the Pacific coast is at Isla Rasa off Baja California (which also supports large numbers of Heermann's Gulls *[Larus heermanni]* and Elegant Terns), where 8,000 to 10,000 pairs nest. In southern California, Royal Terns nest sparingly in Elegant Tern colonies in Los Angeles, Orange, and San Diego Counties.

SIZE: Length 20 in. **ADULT BREEDING:** Mar.–June. Sexes similar. White head and underparts, pearl gray upperparts, short black crest swept back. Thick, straight orange bill. **ADULT NONBREEDING:** June–Mar. Sexes similar. Similar to adult breeding, but black cap (eye is always outside cap) much reduced to show white forehead. **STATUS:** Fairly common local winter visitor south of Morro Bay. Rare local summer resident from Los Angeles County south. **(B*)**

ELEGANT TERN   *Sterna elegans*
(Pl. 92)

In summertime at south San Diego Bay or Bolsa Chica, the constant nasal cry of the Elegant Tern "karREEK, karREEK, karREEK" can be heard as the birds fly back and forth with freshly caught fish to their chicks back at the nesting colonies. These noisy, agile birds execute plunge-dives to catch Northern Anchovies *(Engraulis mordax)* and Pacific Sardines *(Sardinops sagax)*, as they forage in nearshore waters. Their thin orange bills and drooping black crests distinguish them in breeding plumage.

The Elegant Tern has a very limited breeding range, nesting in mainland or island colonies only in southern California and Baja California. Over 90 percent of the world population of Elegant Terns breeds in the company of Heermann's Gulls *(Larus heermanni)* and Royal Terns *(Sterna maxima)* on a tiny island off the east coast of Baja, Isla Rasa. Other colonies are at South San Diego Bay, Bolsa Chica, Port of Los Angeles, and the Colorado River delta.

Elegant Terns are most common on southern California's coast in late summer and early fall, when local breeders are joined

Plate 92. Elegant Tern: adult nonbreeding.

by those dispersing northward from Baja. At this time, the fledged young accompany the adults, which continue to feed the juveniles until they are well into their first year. Family groups may range as far north as Humboldt Bay (see also pages 46, 48).

SIZE: Length 17 in. **ADULT BREEDING:** Mar.–July. Sexes similar. White head and underparts, pearl gray upperparts. Long, shaggy black crest. Slender, slightly decurved orange bill. **ADULT NON-BREEDING:** Aug.–Feb. Sexes similar. Similar to adult breeding, but black cap reduced (eye is always within cap), white forehead. **STATUS:** July–Oct. Common summer visitor. Fairly common local summer resident from Los Angeles County south. Uncommon spring visitor. **(B)**

FORSTER'S TERN   *Sterna forsteri*
(Pl. 93)
Hovering over coastal lagoons and freshwater lakes, harbors, and high-tide pools, a white tern makes a pretty sight against the blue sky. This small tern with a black smudge through its eye plies the air above while scanning the water below for the telltale leap of the anchovy or the smelt. Spotting a fish, the tern halts in midair, changes direction, and makes a shallow dive. Emerging with a

Plate 93. Forster's Tern: adult nonbreeding.

squirming fish, the Forster's Tern devours it on the wing. Our most frequently observed tern, the Forster's can be counted on to put in an appearance at almost any time of year.

Most Forster's Terns we see breed on marshes and lakes in the interior of the United States and Canada, then spend the winter along the coasts. A few Forster's Tern nesting colonies, however, are found along southern California's shores; among them are those at South San Diego Bay, Sweetwater Marsh National Wildlife Refuge near Chula Vista, and Bolsa Chica.

SIZE: Length: 13 in. **ADULT BREEDING:** Mar.–Aug. Sexes similar. White head and underparts, pale gray mantle, black cap. Elongated tail feathers. Reddish legs. Orange bill with black tip. **ADULT NONBREEDING:** Aug.–Feb. Sexes similar. Similar to adult breeding, but lacks dark cap. Dark eye patch. Black bill. **STATUS:** Common migrant and winter visitor. Uncommon to fairly common local resident in Orange and San Diego Counties. **(B)**

LEAST TERN   *Sterna antillarum*
(Pl. 94)

This tiny tern with a unique black-and-white head pattern and a yellow bill returns year after year to traditional nesting sites along southern California's beaches. Choosing a habitat for nesting that

Plate 94. Least Tern: adult breeding.

was once plentiful—deserted beaches backed by a lagoon or river mouth—the Least Tern lays its eggs in the warm, dry sand and fishes in nearby shallow waters.

The pressures of human activity, mostly from beachgoers and housing developments along the shore, have threatened the survival of the subspecies of the Least Tern known as the California Least Tern *(S. a. browni)*. By the 1930s, populations began to decline, and in 1970/71 the bird was listed as endangered by the federal and state governments. California Least Terns are subject to some of the same predators as Clapper Rails *(Rallus longirostris)*: the Red Fox *(Vulpes vulpes)*, Raccoon *(Procyon lotor)*, Opossum *(Didelphis virginiana)*, and human pets. Since this tern was listed, protective measures and constant monitoring have helped it to increase its numbers. Like other tern species, however, reproductive success varies from year to year depending upon weather (storms and high tides cause nest failure), availability of prey, and human disturbance.

Least Tern colonies can be found, though not in great numbers, throughout the United States along interior rivers or along the coasts. The California Least Tern, however, breeds only from San Francisco south to Baja California. The population of Least

Terns in California was 600 pairs in 1973, growing to 2,750 pairs in 1994 and 4,100 pairs in 1998. In 1999, however, there was a 10 percent decline in the breeding population, illustrating just how fragile this species' nesting success can be.

Among the major nesting colonies in southern California are those at South San Diego Bay, Bolsa Chica, Ormond Beach, the Santa Clara River estuary, and the Santa Maria River estuary. By late August, these terns have departed for Central and South America where they spend the winter.

SIZE: Length 9 in. **ADULT BREEDING:** Mar.–Aug. Sexes similar. White forehead, black cap and eye line, white underparts, pearl gray upperparts. Yellow bill with black tip. **STATUS:** Fairly common to uncommon local summer visitor and resident. **(B+, FE, subsp. SE)**

BLACK SKIMMER   *Rynchops niger*
(Pl. 95)

From the beaches of Santa Barbara to the shores of San Diego Bay, a spectacular avian invader has arrived to claim title to a portion of the southern California coast. The bird looks like an oversized tern, wields a mighty red-and-black bill, and never fails to gain the attention and admiration of those who notice birds.

The Black Skimmer is a relative newcomer to southern California. At a time when many bird species have suffered declines, the Black Skimmer is expanding its range. Since it was first recorded in Orange County in 1962, it has undergone a remarkable population explosion from Mexico to southern California. It first nested at the Salton Sea in 1972, then at South San Diego Bay in 1976. At present, the major coastal Black Skimmer nesting colonies—where they nest with various species of terns—are at South San Diego Bay, Upper Newport Bay, Bolsa Chica, and the Port of Los Angeles.

The Black Skimmer's method of feeding is unique: with long wings set motionless, it glides low above a calm sea or a sheltered bay. As it flies, it dips the lower mandible of its bill in the top layer of the water, scooping up small fish. The Black Skimmer's bill is equipped with a lower mandible longer and deeper than the upper mandible, and both are flattened like a knife to slice through the water. When fishing, it feels prey with the lower mandible, clamps down with the upper mandible, and swallows the item on the go.

The Black Skimmer can locate prey by sight and by touch: it is crepuscular (feeding at dawn and dusk) as well as nocturnal. By constricting the pupil of its eye to form a vertical slit during daylight hours, this bird protects its retina from bright light. The retina is dominated by rods—found in many nocturnal birds—

Plate 95. Black Skimmer: adult.

so it can see in low light. At night, when small fish swarm to the surface of the water to feed on invertebrates, the Black Skimmer sees the disturbance made by schooling fish, then scoops them up one by one with its tactile bill.

Black Skimmers are resident along southern shores on the eastern and Gulf coasts of the U.S. On the Pacific coast, they disperse in fall and winter, gathering at beaches from central California to Baja California and Mexico. Black Skimmer numbers appear to be increasing, but they are limited by the availability of suitable nesting sites away from humans (see also page 50).

SIZE: Length 18 in. **ADULT:** Sexes similar. Black wings and upperparts, white underparts. Red legs and feet. Red and black bill with lower mandible elongated. **STATUS:** Uncommon local year-round visitor from Morro Bay south. Fairly common summer resident from Orange County south. **(B)**

AUKS
(Alcidae)

The auks, known as alcids, are pelagic birds with compact bodies and short wings, usually glimpsed from a boat as they whir rapidly away across the surface of the ocean. For most of their lives, alcids forage at sea, only coming to land for nesting on isolated, rocky cliffs.

The alcids comprise murres, guillemots, auklets, puffins, and murrelets. They share adaptations for diving after prey, such as thick, waterproof feathers and the capacity to store oxygen in their body tissues. Alcids are not strong flyers above water, but underwater they are muscular swimmers. Using their wings as propellers and their feet as rudders, alcids pursue fish, squid, and marine invertebrates with great speed beneath the ocean's surface.

The Pigeon Guillemot *(Cepphus columba)* is the alcid most likely to be seen from shore, chiefly north of Point Conception.

PIGEON GUILLEMOT   *Cepphus columba*
(Pl. 96)

From the steep cliffs at Shell Beach or Montana de Oro, a plump, black bird about the size of a pigeon hurtles toward the clear, green water below. It lands with a splash, its bright red legs and feet splayed, then disappears in a dive. Soon, another bird issues with beating wings from a crevice in the rocks and plops down beside its mate in the water. Viewed from above, the two seabirds are striking. They are inky black all over with bright white patches on the wings. Once landed on the water, Pigeon Guillemots float with their necks stretched up and their bills held forward.

The best time to see Pigeon Guillemots is in spring at the entrances to their burrows on sea cliff ledges. During courtship, they display red mouth linings, which match their red legs and feet. A close observer can hear their high-pitched, thin "peeeeeeer"—their mating call—which sounds like the hiss of escaping steam.

Pigeon Guillemots are found only along the Pacific coast from the Bering Strait south to Santa Barbara Island—the species is at the southern tip of its breeding range in our region.

One of the mysteries of Pigeon Guillemot ecology, until recently, was where do they go in winter? From late August through March, the southern California breeding colonies are deserted and Pigeon Guillemots are rarely if ever seen offshore. Recently, observers have established that a postbreeding migration takes place, in which the Pigeon Guillemots that breed along the California coast travel northward to winter in the sheltered marine waters of the San Juan Islands and Puget Sound in Washington, as well as among the islands off British Columbia.

Plate 96. Pigeon Guillemot: adult breeding.

SIZE: Length 13.5 in. **ADULT BREEDING:** Mar.–Sept. Sexes similar. All black; white wing patch crossed with a black wedge. Red legs and feet. Red mouth lining. Black bill. **ADULT NONBREEDING:** Aug.–Feb. Sexes similar. White on head and neck, mottled gray on back, white wing patch with a black wedge. **STATUS:** Fairly common to common local summer resident north of Point Conception and on the Channel Islands south to Santa Barbara Island. **(B)**

BARN OWLS
(Tytonidae)

Barn Owls differ from typical owls and are classified in their own family. They have heart-shaped faces, long legs, and a comblike structure on their middle toes.

Barn Owls are strictly nocturnal, most active an hour after sunset and an hour before dawn. They fly silently through the night, the serrations on the leading edges of their front flight feathers making their wings noiseless. They feed predominantly on voles *(Microtus)* and other small mammals.

Barn Owls have extraordinary hearing. Numerous scientific experiments reveal their ability to locate sounds at night with remarkable accuracy. The feathers of the owl's two facial discs help conduct sound toward its ears. The ears are placed asymmetrically on either side of the head, so the owl can triangulate the location of a mouse by its faintest rustle, depending upon the length of time the sound takes to reach each ear.

Although Barn Owls *(Tyto alba)* are not strictly coastal, they are often observed in coastal lowlands. Another species of owl, the Short-eared Owl *(Asio flammeus),* passes through irregularly along the coast but is not described here.

BARN OWL *Tyto alba*
(Pl. 97)

At night on roads near the coast, the ghostly shape of a white owl flapping across a marsh or open field is occasionally caught in the glare of the car's headlights. This is the Barn Owl, making low quartering flights above the ground searching for prey. Its long, rasping scream creates an eerie sound, but the Barn Owl is simply proclaiming its territory.

Barn Owls nest in cavities in trees, cliffs, caves, and old sheds or barns. Their populations fluctuate from one area to another, depending upon the prey base and available nesting sites. The species readily accepts nest boxes. Along the Santa Barbara and San Luis Obispo County coasts, managers of orchards, vineyards, and other agricultural operations have learned the benefit of having a Barn Owl on the property to control rodents, particularly the

Plate 97.
Barn Owl:
adult.

pocket gopher. As you drive by, notice the Barn Owl nest boxes at the margins of cultivated fields.

In fall and winter, Barn Owls roost in small groups in dense trees during the day. When disturbed, they fly sleepily off to seek another hiding place from the glare of the sun.

SIZE: Length 16 in. **ADULT:** Sexes similar. Pale, tawny upperparts, white underparts. White, heart-shaped facial discs. **STATUS:** Uncommon migrant and winter visitor. Uncommon local resident. **(B)**

KINGFISHERS (Alcedinidae)

Kingfishers are short-necked, short-tailed birds with a big head and crest, making them look top-heavy.

Kingfishers hunt for fish by sight, so they need clear water and a good vantage point. Whether by a stream, a river, a protected harbor, or a backyard fishpond, kingfishers watch for fish from a perch. They hover high above the water before diving with a splash once a fish is targeted.

Kingfishers use their long, pointed bills and short legs and feet to dig a nesting burrow, a 3- to 7-foot tunnel in the side of the bank of a stream or river. At the end of the tunnel, a nesting chamber holds the eggs and young of the kingfisher. To feed the young, the adult birds scramble along the tunnel bearing fish. By the time the nestlings fledge, the chamber is littered with scales and old fish bones.

One species, the Belted Kingfisher *(Ceryle alcyon),* inhabits southern California's coast.

Plate 98. Belted Kingfisher: juvenile (Brad Sillasen).

BELTED KINGFISHER   *Ceryle alcyon*

(Pl. 98)

Perched on a utility cable strung across a coastal lagoon, a chunky, slate blue bird with a shaggy crest and a white collar balances itself against the onshore breeze. As it rocks back and forth, it eyes the water below for the slightest movement. Suddenly, the bird hits the water with a splash, then flies off with a fish held in its bill. Landing atop a post nearby, the Belted Kingfisher attempts to subdue its squirming prey by banging it against the post. Once this is accomplished, the kingfisher swallows the fish head first with a huge gulp.

The Belted Kingfisher always chooses a conspicuous spot from which to fish, whether it is the masthead of a sailboat in a harbor or a prominent piece of driftwood in a salt marsh. Often, the kingfisher warns of its approach with a staccato, rattling call, heard long before the bird appears.

The female Belted Kingfisher is larger and more colorful than the male. Across her breast, she has a blue belt, as well as a chestnut one. The male has only a blue belt.

SIZE: Length 13 in. **JUVENILE:** Dark breastband, smudge of chestnut on flanks. **ADULT MALE:** Slate blue above, white collar and underparts, blue belt across chest. **ADULT FEMALE:** Slate blue above, white collar and underparts. Blue and chestnut belts across chest, chestnut on flanks. **STATUS:** Fairly common migrant and winter visitor. Rare local resident. **(B*)**

TYRANT FLYCATCHERS (Tyrannidae)

This family's Latin name comes from the Greek *tyrannos* meaning "ruler," which refers to the aggressive, noisy behavior of some of the larger flycatchers.

Flycatchers sit erect on a perch, watching for insects. On spotting prey, they fly out and grab it in midair with a snap of the bill. Although it appears slender, the bill is actually flattened sideways, increasing the bird's ability to successfully nab an insect on the wing.

Flycatchers belong to the suboscines, which differ from true songbirds, or oscines, in their vocal structure. Flycatcher vocalizations are simple syllables, lacking the tuneful repertoire of many songbirds.

The Black Phoebe *(Sayornis nigricans)* and Say's Phoebe *(S. saya)* frequent the southern California coast.

BLACK PHOEBE   *Sayornis nigricans*
(Pl. 99)

A small black bird with a peaked head dips and pumps its tail as it sits on the guardrail of a bridge or watches for flies from a pipe crossing a ditch. Its gleaming white belly, which comes up to border the black chest and flanks in an inverted V, gives somewhat the appearance of a cutaway waistcoat.

Plate 99.
Black Phoebe:
adult.

Black Phoebes frequent coastal cliffs, ponds, and riverbanks, as well as parks and gardens: anywhere they can find water (which draws insects) and mud (for nest building). They sit on a low perch, such as a water faucet in the middle of a lawn or a rock beside a stream, then hawk insects by chasing them in flight.

The species has adapted so well to human haunts that it is expanding its range. Rather than being limited to the boulders and rock faces found in nature, the Black Phoebe has learned to attach its mud and straw nest to the walls and eaves of buildings and porches, or under bridges and other structures, provided they are close to water.

The Black Phoebe is a bird found only in California and the Southwest and is mostly nonmigratory within its breeding range.

SIZE: Length 7 in. **ADULT:** Sexes similar. Black head, chest, and back, white underparts. **STATUS:** Common resident. **(B)**

SAY'S PHOEBE *Sayornis saya*
(Pl. 100)

On a blustery winter day when shorebirds are huddled out of sight and clouds hang heavy on the salt marsh, a small flycatcher hovers low above the pickleweed. Buffeted by the wind, the flycatcher eventually alights on a post and sits, dipping its black tail and showing a pale, cinnamon-colored belly.

Plate 100.
Say's Phoebe:
adult.

The Say's Phoebe—named to commemorate Thomas Say, a nineteenth-century entomologist who explored the West—is closely related to the Black Phoebe *(S. nigricans)*.

Say's Phoebes range throughout the West in North America: from Alaska all the way to Mexico, and east to the Great Plains. After nesting in arid canyons and sagebrush flats in the interior, Say's Phoebes migrate to the southern California coast to spend the fall and winter, where they forage over coastal lowlands in open fields, scrubby areas, and salt marshes.

SIZE: Length 7.5 in. **ADULT:** Sexes similar. Gray brown upperparts and breast, contrasting with black tail. Pale orange belly. **STATUS:** Fairly common fall and winter visitor.

CROWS and RAVENS (Corvidae)

Crows and ravens belong to a large family of birds, the corvids. Reportedly some of the most intelligent of all birds, crows and ravens have performed well in various experiments. They are, however, extremely difficult to capture and band, so anecdotal evidence of these birds' intelligence may be somewhat exaggerated. That said, several examples of corvid intelligence are impressive. When given a piece of meat on a string dangling from a perch, a raven, who had never seen string, used its foot to pull up the string to get the food up to the perch so it could be eaten. A species of crow in New Caledonia makes a tool by stripping a twig of leaves and bark and using the curved, pointed end to extract insects from holes in trees.

Crows and ravens are large, black birds with thick black bills. They have strong legs, feet, and toes, which help them feed on a variety of prey. Being wary and observant has helped crows and ravens figure out unfamiliar situations and learn to exploit them for food.

The American Crow *(Corvus brachyrhynchos)* and Common Raven *(C. corax)* are two species of corvid that frequent the southern California coast.

AMERICAN CROW   *Corvus brachyrhynchos*

(Pl. 101)

At twilight, a flock of several hundred black birds with rounded tails and broad wings flies across an evening sky. Their grating "caw, caw, caw" call lends an ominous feeling, as the dark forms pass on their way to roost in a clump of cypress trees. Reaching the trees, the birds make a ruckus of "cawing" noises. They dive and swoop at a brown lump perched on one of the branches. A

Plate 101. American Crow: adult.

hawk has blundered into their nightly roosting spot, and the flock will not rest until it has driven the bird of prey away.

The ubiquitous American Crow, known as just "a crow" to most people, was historically found only in rural areas. In the nineteenth and early twentieth centuries, the crow was persecuted and shot in the countryside, making the bird avoid people. In recent decades, however, the species has taken refuge in towns and suburban landscapes, where people do not shoot at them. Although American Crow populations have steadily increased, it may also be that the public is conscious of their presence now, because the birds are more visible around urban settings than ever before.

At the coast, American Crows scavenge on dead fish and marine mammals that wash up on the beach. They drop clams from a height and descend to eat the meat, investigate tide pools to grab crabs or other creatures, and watch for roadkills along highways. In addition, American Crows (as well as Common Ravens *[C. corax]*) have been seen to plunder Snowy Plover *(Charadrius alexandrinus)* and Least Tern *(Sterna antillarum)* eggs at their nesting colonies on southern California beaches.

The American Crow is widely distributed across North America. In winter, crows breeding in more northerly areas move south, where they join resident crows. At this season, large flocks of crows form communal roosts in tall trees, where they gather every evening.

SIZE: Length 17.5 in. **ADULT:** Sexes similar. All black with glossy sheen. **STATUS:** Common resident. **(B)**

COMMON RAVEN   *Corvus corax*
(Pl. 102)

Where sandstone bluffs fall directly to the water along our coast, a pair of large black birds with wedge-shaped tails soars in circles. Their steady flight and all-black appearance coupled with a hoarse, croaking "cur-ruk" call signal the approach of a couple of

Plate 102. Common Raven: adult.

Common Ravens. They alight at their nest site, a scraggly pile of sticks protruding from a ledge on the cliffs.

Common Ravens are bold, fascinating birds. Skilled aerialists, they ride the wind currents near coastal bluffs or dive and roll in the sky high above suburban communities. Ravens' wariness and resourcefulness help them survive along our coast: they scavenge on carcasses washed up on the beach, raid the young of breeding seabirds, and drop sea urchins from a height onto rocky shores. They have also discovered how to feed at municipal landfills.

Ravens can be told from crows by their larger size, diamond- or wedge-shaped tail in flight, larger bill, and elongated throat feathers.

Common Ravens are one of the most widespread birds in the world, found almost everywhere except the Neotropics. Along the southern California coast, the Common Raven has increased its population over recent years. Interestingly, it is quite rare along the San Luis Obispo County portion of the coast.

SIZE: Length 24 in. **ADULT:** Sexes similar. All black. Elongated throat feathers. Large, long black bill. **STATUS:** Uncommon to fairly common resident. Rare winter visitor north of Point Conception. **(B)**

SWALLOWS and MARTINS (Hirundinidae)

Swallows and martins are small birds that swoop and glide in the air, their mouths held wide open to take in a wealth of flying insects. When swallows come to rest, they perch on wires, reeds, or exposed twigs, seldom in the leafy portion of trees.

Birders occasionally confuse swallows with swifts, which are in a separate family. Although both fly about feeding on insects, their wing structure is very different. A swift's wing is thin and bowed back in an arc. The swallow wing is more like that of an airplane, thicker where it is attached to the body, tapering to a point at the end.

From their winter homes in Central and South America, swallows migrate great distances every spring to breeding grounds here and as far north as Alaska. In fall, they repeat this amazing journey on the return trip south.

Three species of swallows are common along the southern California coast: the Tree Swallow *(Tachycineta bicolor)*, Cliff Swallow *(Petrochelidon pyrrhonota)*, and Barn Swallow *(Hirundo rustica)*.

TREE SWALLOW   *Tachycineta bicolor*
(Pl. 103)

On a drizzly day in February, when spring seems far away, a small group of swallows with dark, greenish blue backs and gleaming white underparts comes dashing low in zigzag flight over our estuaries and salt marshes. These Tree Swallows are in the forefront of spring bird migration and one of the earliest of the swallows to arrive along our coast.

Plate 103. Tree Swallow: adult male.

Tree Swallows compete for nesting sites in cavities in rotting trees, typically near water. To prepare for nesting, the species arrives very early from its wintering grounds, males defend their nest sites aggressively, and first-year females forego nesting altogether. Abandoned woodpecker holes were formerly sought, but with the demise of most swampy areas and the removal of trees for agriculture, the species was close to being extirpated from our region. However, dedicated volunteers began placing nest boxes on poles several feet apart at wastewater treatment ponds and other freshwater wetlands. The swallows readily accepted the boxes for nesting, and Tree Swallow populations have begun to recover substantially over the past decade.

Where riparian thickets still grow near the coast, such as at Oso Flaco Lake in San Luis Obispo County, Tree Swallows nest in old woodpecker holes in the mature willows.

SIZE: Length 5.75 in. **ADULT MALE:** Blue green back, white underparts with white coming up to eye, but not above it. Very slightly forked tail. **ADULT FEMALE:** Duller version of adult male; some are all brown on back. **STATUS:** Common spring migrant, uncommon fall migrant. Uncommon summer resident, uncommon local winter visitor. **(B)**

CLIFF SWALLOW   *Petrochelidon pyrrhonota*
(Pl. 104)

Under freeway bridges or building eaves, many gourd-shaped mud nests cling to the smooth, vertical walls. Composed of hundreds of little beakfuls of mud, the nests have been constructed by members of a swallow colony. The air is thick with birds, as they fly back and forth seeking insects to feed the nestlings. When one of the swallows clings to the narrow neck of its mud nest before entering, it shows a square tail with a pale, buff-colored rump.

The Cliff Swallow is an abundant summer resident along the southern California coast. In addition, it has expanded its range throughout North America, because the building of human-made structures—highway bridges, culverts—has replaced the rock ledges and steep cliffs of the mountainous West that this species historically used.

Cliff Swallows nest in exceptionally large colonies, often comprising many hundreds of birds. They can recognize the voices of

Plate 104. Cliff Swallow: adult on nest.

their own nestlings, and they learn about food sources from watching the success of other residents of the colony. Female Cliff Swallows parasitize neighbors' nests by laying eggs there or by moving eggs from their own nests into others'.

Like the Barn Swallow *(Hirundo rustica)* to which it is closely related, the Cliff Swallow migrates south every fall from its breeding range in North America, through the Central American isthmus, to winter east of the Andes from southern Brazil through Argentina.

SIZE: Length 5.5 in. **ADULT:** Sexes similar. Blue black crown and back, white forehead, rufous throat, white underparts. Pale, buff-colored rump. **STATUS:** Common migrant and summer resident. **(B)**

BARN SWALLOW *Hirundo rustica*
(Pl. 105)

A swallow with a deeply forked tail, rusty underparts, and dark blue back flies out from under the pilings of a wharf or a boat dock near the waterfront and veers off just inches above the water. The Barn Swallow, our only swallow with a deeply forked tail, is easy to recognize.

Barn Swallows fly in swooping arcs over coastal lowlands in pursuit of flying insects. This species' elongated outer tail feathers

Plate 105. Barn Swallow: adult.

give them more maneuverability than other swallows. Indeed, researchers of European populations of the Barn Swallow have measured a 10 percent (.5 in.) increase in the length of the male's outer tail feathers over the last two decades. The extra length of the male's tail advertises his healthy condition and good gene pool to prospective females: a dramatic example of evolutionary change over a short period of time.

Barn Swallows originally nested in caves but have now completely switched to human-made structures. They place their cup-shaped mud nests on beams in barns and outbuildings. Along the coast, they nest under wooden bridges, boardwalks, and piers. Their mud nests are open at the top, not enclosed like those of the Cliff Swallow *(Petrochelidon pyrrhonota)*, so they require some sort of overhang for protection.

Barn Swallows are the most abundant swallow in the world. In our hemisphere, they migrate between North and South America annually. By September, they depart the southern California coast, returning from their long-distance journey the following spring.

SIZE: Length 6.75 in. **ADULT MALE:** Bluish black upperparts, chestnut throat. Buff-colored underparts, deeply forked tail. **ADULT FE-MALE:** Similar to male but whitish underparts, shorter tail. **STATUS:** Fairly common migrant and summer visitor. Fairly common local resident from Ventura County north. Rare winter visitor. **(B)**

WRENS
(Troglodytidae)

Wrens are a family of small brown birds with short tails, often held cocked up at an angle. Wrens can be difficult to spot, because they sneak through underbrush, woodpiles, and marsh vegetation, looking for insects and spiders. Their thin, slightly decurved bills are perfect for getting bugs out of corners and crevices. Wrens are vociferous in springtime; the remainder of the year they give short, scolding notes.

The Marsh Wren *(Cistothorus palustris)* frequents wetlands along the southern California coast. Another wren species, the Bewick's Wren *(Thryomanes bewickii),* is occasionally found in coastal sage scrub but is not included here.

MARSH WREN   *Cistothorus palustris*
(Pl. 106)

On a spring morning at Upper Newport Bay or San Joaquin Marsh, the tules rustle and bend as a bird flits secretively from one clump to another. A sharp scolding note dominates the marsh, followed by a loud, rollicking song. But the bird remains hidden. At last, a male Marsh Wren inches its way to the top of one of the tules. Holding tightly to the swaying reed, a wren with a dark crown and a white eyebrow throws back its head, bursting forth with a series of buzzy trills and gurgling notes in phrase after phrase.

Ornithologists studying the Marsh Wren found that the species has a polygynous mating system: 50 percent of the males in most populations mate with two or more females. Evidently, the complexity of the males' songs helps them acquire a better territory within the marsh and attracts more females. Interestingly, male Marsh Wrens in the western portion of their North American range have a larger vocal repertoire than eastern males. The two populations have inherited different abilities to learn songs. During the early phase of their development, western males learn 50 to 200 types of songs. In spring, they have singing duels with neighboring males, engaging in competitive countersinging to impress the females and mark their territories.

Plate 106. Marsh Wren: adult.

Marsh Wrens build nests by twisting reed stems into a globe, which they strap to live bulrushes or cattails. Like other wren species, the males may construct half a dozen dummy nests, built to impress the female or as a decoy to fool predators, but not used to shelter a brood.

Marsh Wrens are fairly common as winter visitors to coastal wetlands, frequently encountered in salt marshes at this season. For breeding, they prefer freshwater marshes: south of Point Conception, breeding colonies are few—notably at Upper Newport Bay, San Joaquin Marsh, and certain of the San Diego County lagoons.

SIZE: Length 5 in. **ADULT:** Sexes similar. Dark crown, white eyebrow. Black patch with white streaks on back, cinnamon lower back and rump, whitish underparts. **STATUS:** Fairly common fall migrant and winter visitor. Fairly common resident north of Point Conception. Uncommon local resident elsewhere. **(B)**

STARLINGS and MYNAS (Sturnidae)

Starlings belong to an Old World family of birds found in Europe, North Africa, and Asia.

With their chunky bodies, short legs, and sharp, pointed bills, starlings have a distinctive shape. They use their strong bills to pick at the ground, dislodging soil invertebrates and seeds.

The European Starling *(Sturnus vulgaris)* frequents the southern California coast.

EUROPEAN STARLING *Sturnus vulgaris*
(Pl. 107)

A group of dark birds, each with a very short tail and a waddling gait, walks along a sandy beach near the kelp line or on the grass at a beachside park. Pausing in their search, the birds vigorously peck at the ground for tidbits, then move on.

When flushed, a flock of European Starlings flies in a tight group, twisting and turning in unison. They alight in trees or on utility wires, where the males utter a variety of odd squeaks, chirps, croaks, and whistles—the starling's song. They sometimes imitate the calls of other species.

European Starlings were introduced to North America in 1890/91—when 100 individuals were released in New York City—and have since spread throughout the continent. They are now one of the most abundant birds in North America. Some consider the species a nuisance, because its cavity-nesting habits rob many native species, such as bluebirds and woodpeckers, of potential nest sites.

The European Starling has adapted well to a number of habitats, particularly urban surroundings. Along the coast, it can be found at beaches, grassy playing fields, and parking lots. In fall and winter, European Starlings roost at freshwater marshes or groves of trees along with large flocks of Red-winged and Brewer's Blackbirds *(Agelaius phoeniceus* and *Euphagus cyanocephalus).*

European Starlings are named for their fall plumage, in which they are spotted with many white dots, "little stars." By spring, as their bright feather edgings wear off, the starling in breed-

Plate 107. European Starling: adult breeding.

ing plumage becomes an iridescent black. Its bill turns a bright yellow.

SIZE: Length 8.5 in. **ADULT BREEDING:** Jan.–Aug. Sexes similar. Glossy black plumage with faint spots. Pinkish legs. Yellow bill. **ADULT NONBREEDING:** Sept.–Feb. Sexes similar. Blackish plumage speckled with white dots. Dark bill. **STATUS:** Common resident. **(B)**

WAGTAILS and PIPITS (Motacillidae)

Pipits and wagtails, common in Africa and Eurasia, are represented by only a few species in North America.

Because most pipits are brown with streaked breasts and spend time on the ground, beginning observers might confuse them with sparrows. Unlike sparrows, pipits often pump their tails up and down as they walk, moving their heads forward with each step. They forage in flocks through open areas gleaning insects and seeds with short, thin bills. On the shore, they search piles of kelp to uncover insects.

The American Pipit *(Anthus rubescens)* is the only pipit to regularly visit southern California's shores.

AMERICAN PIPIT
Anthus rubescens
(Pl. 108)

From time to time along our beaches in winter, a brown-backed bird with a brown-streaked pale breast walks across the dry sand to inspect the sea wrack washed up by the night's high tide. As it walks, it holds its head up, and, when it stops to rest, bobs its tail slightly. Occasionally, the pipit jumps into the air to grab an insect, spreading its tail, showing white outer tail feathers. The slender, long-tailed shape of the American Pipit (formerly Water Pipit), together with its tiny bill, should eliminate any confusion with nearby shorebirds on the beach.

Plate 108. American Pipit: adult nonbreeding.

In their fall migration and during the winter, American Pipits gather in restless flocks, flying over open fields. They call often—their distinctive, high-pitched "sip-sip" (supposedly sounding like "pipit") carrying in two thin syllables across the sky.

Although American Pipits overwinter at low elevations, in spring they leave for their breeding grounds on Arctic tundra and at high elevations in western mountains.

SIZE: Length 6.5 in. **ADULT NONBREEDING:** Aug.–Mar. Sexes similar. Faintly streaked brown back. Whitish underparts tinged with buff, heavy brown streaking concentrated across the breast. Dusky legs. **STATUS:** Common migrant and winter visitor.

WOOD-WARBLERS (Parulidae)

Warblers are the jewels of the bird world. Vibrantly colored and constantly active, these tiny birds flit from tree to bush. Never still for long, they search the leaves and twigs to feed their appetites for insects and spiders.

Birders like warblers for their beautiful plumages and intricate songs. Warblers are a challenge to identify, especially the dull females and immatures in nonbreeding plumage.

The Common Yellowthroat *(Geothlypis trichas)* is a common warbler along the southern California coast. Another warbler occasionally found near the coast in fall and winter, the Yellow-rumped Warbler *(Dendroica coronata),* is not included here.

COMMON YELLOWTHROAT   *Geothlypis trichas*
(Pl. 109)

From tules bordering a freshwater pond or from low shrubbery near a park bench, an abrupt, scolding "chek, chek" warns that a bird is skulking nearby. A little squeaking on the part of the birder usually results in a glimpse of a small bird with a bright yellow throat and a black mask. At length, the male Common Yellowthroat pops into view to investigate the intruder.

Common Yellowthroats are one of America's most widespread warblers. They are chunkier and more sluggish than most warblers, however, and are found in dense, low vegetation, not high in the treetops. Yellowthroats often live in marshy areas, where they can be glimpsed foraging among the cattails, similar to a Marsh Wren *(Cistothorus palustris).*

The male Common Yellowthroat has an easily recognized song, which he repeats loudly over and over from a perch in the

Plate 109. Common Yellowthroat: adult male.

breeding season: "WHICH-e-ty, WHICH-e-ty, WHICH-e-ty." In one scientific study, researchers noted that males sang this perching song phrase more softly and less frequently after they had paired with a female.

Once seen well, adult male Common Yellowthroats are easy to identify; however, the females are brownish, with pale yellow throats and pale yellow undertail coverts and no black mask.

Most of our breeding Common Yellowthroats are sedentary, residing in southern California year-round. The population is augmented in winter by other Common Yellowthroats that have migrated here from northern areas.

SIZE: Length 5 in. **ADULT MALE:** Brownish above, gray brown belly. Bright yellow throat and breast, black mask across forehead bordered by white. **ADULT FEMALE:** Brownish above, pale gray belly, yellow throat and undertail coverts. **STATUS:** Common fall migrant and winter visitor. Fairly common resident. **(B)**

SPARROWS (Emberizidae)

Sparrows spend much of their time on or near the ground. They frequent suburban parks and gardens, coastal sage scrub, streambeds, and weedy patches along coastal greenbelts and bike paths. Sparrows' conical bills are suited for cracking seeds, the size of the bill dictating the size of the seeds the species can husk. During the breeding season when sparrows raise young, many species switch to an insect diet. The extra protein provided by insects nourishes the nestlings and enables the birds to raise more than one brood. Once winter comes and insect life diminishes, sparrows return to their seed-eating habits.

Several species of sparrow are found along the southern California coast, including Spotted Towhee *(Pipilo maculatus)*, California Towhee *(P. crissalis)*, Savannah Sparrow *(Passerculus sandwichensis)*, Song Sparrow *(Melospiza melodia)*, White-crowned Sparrow *(Zonotrichia leucophrys)*, and Golden-crowned Sparrow *(Z. atricapilla)*. The Belding's Savannah Sparrow *(P. s. beldingi)*, a subspecies of the Savannah Sparrow, is also included.

SPOTTED TOWHEE	*Pipilo maculatus*
CALIFORNIA TOWHEE	*P. crissalis*

(Pls. 110, 111)

In brushy tangles of wild blackberry *(Rubus ursinus)* and poison oak *(Toxicodendron diversilobum)* bordering patches of coastal sage scrub, a colorful bird with a black head, rusty flanks, and white belly hops to the topmost branch of a shrub and gives a long buzzy trill: "chrrrrrrr."

The **SPOTTED TOWHEE**, formerly Rufous-sided Towhee, keeps hidden in dense thickets and is often difficult to see. Like its relative the California Towhee, the Spotted Towhee uses the double-scratch method of feeding on the ground: with feet together, it hops backward and forward to dislodge seeds and insects by disturbing the leaf litter.

In contrast, the **CALIFORNIA TOWHEE** takes no pains to stay hidden, being one of the most familiar southern California birds. Usually in full view, the California Towhee searches for weed seeds

Plate 110.
Spotted
Towhee:
adult male.

Plate 111.
California
Towhee:
adult.

at the base of bushes in coastal sage scrub or around the edges of lawns in city parks. A large, overall brown sparrow, the California Towhee, formerly Brown Towhee, is a true ground dweller. In spring, however, it may mount to the top of a small tree to sing its song, a monotonous "chink, chink, chink-chink-chink-chink." The California Towhee has adapted well to urbanization and is increasing its population in our region.

Both Spotted and California Towhees are nonmigratory, permanent residents along the southern California coast.

SPOTTED TOWHEE SIZE: Length 8.5 in. **ADULT MALE:** Black head, chest, and back. White spots on wings, rufous flanks, white belly. **ADULT FEMALE:** Similar to adult male, but head, chest, and back are blackish brown. **STATUS:** Common resident. **(B)**

CALIFORNIA TOWHEE SIZE: Length 9 in. **ADULT:** Sexes similar. Uniform gray brown, buff-colored chin, rufous undertail coverts. **STATUS:** Common resident. **(B)**

SAVANNAH SPARROW   *Passerculus sandwichensis*
(Pl. 112)

A small sparrow with a short, slightly notched tail flushes from the dry weeds by the side of the road and quickly flies several yards away to a fence wire. For an instant it perches there, long enough for an observer to notice its finely streaked breast, yellowish eyebrow, and pink legs and feet. Immediately, the sparrow dives into the surrounding weeds, giving a high, thin "tsip" note as it disappears.

This is the Savannah Sparrow, a bird of open grasslands and cultivated fields. It nests in northern and interior areas of North America, then comes to the coast to spend the fall and winter. A Savannah Sparrow typically sneaks through the grass while foraging for seeds. When disturbed, it flies to the top of a weed stalk or small shrub to briefly survey the surroundings.

The best way to separate the Savannah Sparrow from other sparrows is by the whitish stripe down the middle of its crown. The Savannah Sparrow can be told from the Song Sparrow *(Melospiza melodia)* by its thinner, paler breast streaks. Although the breast streaks sometimes form a dark spot in the center of the chest, it is not as bold as that of the Song Sparrow.

Plate 112. Savannah Sparrow: adult.

SIZE: Length 5.5 in. **ADULT:** Sexes similar. Pale brown above with dark streaks, whitish median crown stripe. Often shows a yellowish eyebrow. White underparts with fine, dark streaking, sometimes forming a spot. Pink legs. **STATUS:** Common migrant and winter visitor.

BELDING'S SAVANNAH SPARROW

(Pl. 113)

Passerculus sandwichensis beldingi

In spring and summer, after the wintering shorebirds and ducks have flown north, southern California salt marshes south of Point Conception are the domain of the only two birds that regularly breed there: the Light-footed Clapper Rail *(Rallus longirostris levipes)* and the Belding's subspecies of the Savannah Sparrow.

A darker and more heavily streaked version of the Savannah Sparrow *(P. sandwichensis)*, the Belding's subspecies is found only in salt marshes from Devereux Slough (Coal Oil Point Reserve) south to Baja California, where it lives out its entire life in the midst of the low-growing pickleweed.

In 1974, the Belding's Savannah Sparrow was listed as endangered by the state of California, as a result of rapidly dwindling salt marsh habitat along the southern California coast. Since then,

Plate 113. Belding's Savannah Sparrow: adult.

statewide surveys have shown that the breeding populations of this little sparrow can hold their own in the larger marshes, but they tend to decline in smaller wetlands. In 1996, one-fourth of the breeding populations surveyed were composed of 20 or fewer pairs. The largest numbers were at Mugu Lagoon (400 pairs), Upper Newport Bay (252 pairs), and Tijuana Slough (250 pairs). In 1999, the status of the species was listed as stable to declining.

SIZE: Length 5.5 in. **ADULT:** Sexes similar. A darker, more heavily streaked version of the Savannah Sparrow. **STATUS:** Fairly common local resident in salt marshes south of Point Conception. **(B*, SE)**

SONG SPARROW   *Melospiza melodia*
(Pl. 114)

A dark brown sparrow with a streaked breast flushes from the grasses near a coastal creek, then flies across the trail in jerky flight with tail pumping. This is the Song Sparrow—a sort of template of sparrowdom—and a good starting point for a course of sparrow study. Year round, it is our commonest sparrow, inhabiting brushy areas near creeks, ponds, and coastal sage scrub bordering tidal marshes.

When the Song Sparrow sits at the top of a shrub to announce

Plate 114. Song Sparrow: adult.

its territory to the world, notice the pattern on either side of the throat: a pair of tan and black "mustache" stripes. The sparrow's white underparts are streaked with dark, thick lines, which sometimes form a black dot or "stickpin" in the center of the breast.

The Song Sparrow's song is easy to recognize. It begins with three clear introductory notes, then a higher note which is held briefly, followed by a series of descending trills. Henry David Thoreau interpreted it as "Maids! Maids! Maids! Hang up your teakettle-ettle-ettle."

Song Sparrows breed from Alaska to central Mexico throughout North America and are migratory in most parts of their range. In southern California, however, many Song Sparrows are year-round residents, defending their territories at all seasons. Interestingly, their population is especially dense in human residential developments near coastal sage scrub along the coast.

SIZE: Length 6.25 in. **ADULT:** Sexes similar. Reddish brown wings and tail. Gray face with tan and black "mustache" stripes. White below with heavy black streaks on breast forming central dark spot. **STATUS:** Common migrant and winter visitor. Fairly common resident. **(B)**

WHITE-CROWNED SPARROW   *Zonotrichia leucophrys*
(Pl. 115)

In fall and winter, at the base of bushes in a park, or along a trail through coastal sage scrub, or beneath a garden bird feeder, several sparrows with pale gray breasts and striking crown patterns of alternating black-and-white stripes scratch for seeds on the ground. The sparrows forage near the shelter of a thick hedge or tangle of weeds; if startled, they dive into the undergrowth and sit quietly. As soon as the birds feel safe, they fly to the ground again, one by one, to resume feeding.

Plate 115. White-crowned Sparrow: adult.

These are White-crowned Sparrows, which breed over much of the northern tier of North America and winter in the southern portion. The species comprises five distinct subspecies, many of which have different behaviors: some are long-distance migrants, some are short-distance migrants, and some are permanent residents. Most White-crowned Sparrow subspecies can be recognized by subtle differences in their songs and plumages, a boon to researchers who have studied these birds extensively.

Wintering flocks in southern California are generally made up of White-crowned Sparrows that come from Alaska *(Z. l. gambelii),* and from the Pacific Northwest *(Z. l. pugetensis);* however, north of Point Conception, the Nuttall's White-crowned Sparrow *(Z. l. nuttalli)* is a permanent resident, breeding in a narrow band along the California coast from Point Conception to northwest California.

Many important aspects of White-crowned Sparrows' migratory and breeding ecology have been discovered. For example, resident Nuttall's White-crowned Sparrows maintain year-round territories. The juveniles (hatched in spring) learn songs in the first months after fledging, then refine their singing skills to begin to stake out territories in September and October. They learn the songs of adult Nuttall's White-crowneds, not the songs of the other wintering subspecies (which vocalize in winter) to which they are constantly exposed.

Migratory subspecies of White-crowned Sparrows return to the same wintering sites year after year. In coastal California, researchers have observed that these wintering flocks establish a dominance hierarchy in the following pattern: adult males, adult females, immature males, immature females. It appears that group hierarchy is strongest in foraging areas closest to cover, because that location is perceived by the birds to be more valuable or safer. The farther away from cover a group of White-crowneds feeds, the less aggressive are the interactions within the flock, because they are not competing for the prime spots near good cover.

SIZE: Length 7 in. **IMMATURE:** Sexes similar. Similar to adult, but crown striped with rufous and buff instead of black and white. **ADULT:** Sexes similar. Brownish back and wings, black-and-white crown stripes. Gray face, neck, and breast. Pale bill. **STATUS:** Common migrant and winter visitor. Common permanent resident north of Point Conception only. **(B*)**

GOLDEN-CROWNED SPARROW   *Zonotrichia atricapilla*
(Pl. 116)

Scanning a wintering group of White-crowned Sparrows *(Z. leu-cophrys)* as they feed on a lawn, a birder notices a bigger, browner sparrow skulking near the edge. The bird's head pattern shows a faint tinge of yellow on the front of its crown, bordered by two dark crown stripes.

The Golden-crowned Sparrow, larger and more reclusive than its White-crowned relative, arrives on our coast in late September and October and stays through April. During their stay here, Golden-crowned Sparrows inhabit coastal sage scrub, riparian

Plate 116. Golden-crowned Sparrow: adult breeding.

thickets, and parks and gardens, where they forage on the ground by pecking and scratching.

The species has migrated a fair distance from its nesting range, which lies along the north Pacific coast from Alaska to British Columbia. There, it nests in shrubby areas from the coast to treeline. During the gold rush at the end of the nineteenth century, miners in Yukon Territory would comment on the mournful, descending three-note song of the Golden-crowned Sparrow, paraphrasing it as "I'm so tired!" or "No gold here." A more recent rendition is referred to as "Three blind mice."

SIZE: Length 7.25 in. **ADULT BREEDING:** Sexes similar. Similar to adult nonbreeding, but bright gold forecrown bordered by thick black crown stripes. **ADULT NONBREEDING:** Sexes similar. Gray face and neck, gray brown chest. Yellow tinge on forecrown bordered by two faint dark crown stripes. Dusky bill. **STATUS:** Common migrant and winter visitor.

BLACKBIRDS and ORIOLES (Icteridae)

The brilliantly colored orioles, the tan-and-yellow meadowlarks, and several kinds of blackbirds comprise the icterids. Except for the meadowlarks, the males of this family are more colorful than the females.

All icterids have long, sharply pointed bills. Most have strong legs and feet for walking on the ground.

Along the California coast the following icterid species can be seen: the Red-winged Blackbird *(Agelaius phoeniceus)*, Western Meadowlark *(Sturnella neglecta)*, Brewer's Blackbird *(Euphagus cyanocephalus)*, and Hooded Oriole *(Icterus cucullatus)*.

RED-WINGED BLACKBIRD   *Agelaius phoeniceus*
(Pl. 117)

On a spring day, when fields along the coast are covered in tall yellow-flowered mustard, an all-black bird with brilliant red wing patches grasps the end of one of the tall weeds and sings over and over "conk-a-REEE?" As it sings, the bird flares its red epaulets and lowers and spreads its black tail to impress a female.

The Red-winged Blackbird is reportedly the most abundant bird in North America. Ornithological research on the species is extensive; many elements of Red-winged Blackbirds' social organization and feeding habits have been studied.

Typically, we think of Red-winged Blackbirds as nesting in freshwater marshes, but they also use fields of tall weeds. Their

Plate 117.
Red-winged
Blackbird:
adult male.

nesting colonies are highly social. The majority of the males are polygynous, with up to 15 females observed on the territory of one male. This has the advantage of allowing several females to prosper from an individual male's superior territory, while at the same time gaining protection from predators by the presence of surrounding females on nests.

Both male and female Red-winged Blackbirds have a variety of songs and calls. Both respond to predators by alarm calls and mobbing. Switching from one alarm call to another, males act as sentinels to provide information about predators. In one study, scientists established that this early warning system of male Red-winged Blackbirds prompted females to leave their nests sooner—when a human predator approached—than if the females had not been alerted by the male's calls.

After the breeding season, large flocks of Red-winged Blackbirds find grains and seeds by searching agricultural fields and livestock pens. In fall and winter, they join mixed-species flocks in nighttime roosts in tule marshes near the coast, their raucous calls creating a din at sunset.

SIZE: Length 8.75 in. **ADULT MALE:** All black body. Reddish shoulder patch bordered by pale yellow. Black bill. **ADULT FEMALE:** Dark brown above, pale brownish streaks below. **STATUS:** Common resident. **(B)**

WESTERN MEADOWLARK   *Sturnella neglecta*
(Pl. 118)

In spring, where pastures and open fields border California Route 1 along the San Simeon coast, every other fencepost seems to sport a bird showing a bright yellow breast with a black V-shaped bib. From this perch, the male Western Meadowlark sends forth his glorious, melodic song over the hillsides. As his exuberance mounts, he flies high into the sky, still singing his flutelike, bubbling song.

Plate 118. Western Meadowlark: adult breeding.

When groups of Western Meadowlarks walk on grassy fields, their yellow breasts are hidden due to the cryptic coloration of tan-and-black streaks on their backs. But when they fly, the birds have obvious white outer tail feathers.

The Western Meadowlark is abundant throughout most portions of the western United States, where it inhabits native grasslands, alfalfa fields, orchards, and other open areas. The males are excellent singers. Ornithologists have observed that males with a more varied vocal repertoire have larger territories, pair earlier, and have greater nesting success.

Western Meadowlarks feed on the ground, stabbing their sharp bills into the soil, then opening them once they are inserted into the ground. This method of feeding, called gaping (used by other icterids, as well as European Starlings *[Sturnus vulgaris]*), allows them to forage for crickets, worms, and beetles in spring and summer. In fall and winter, they subsist on seeds and grain.

For nesting, Western Meadowlarks require grasslands, a habitat that has been greatly diminished along the southern California coast over the last half century. In fall and winter, they frequent low-growing vegetation of almost any kind, such as salt marshes and cultivated fields.

SIZE: Length 9.5 in. **ADULT BREEDING:** Feb.–Aug. Sexes similar. Upperparts pale buff streaked with brown. Yellow throat, breast, and belly, with black crescent on chest. White outer tail feathers. **ADULT NONBREEDING:** Sept.–Jan. Sexes similar. Similar to adult breeding, but paler with less contrast. **STATUS:** Common resident north of Point Conception. Fairly common migrant and winter visitor elsewhere. **(B)**

BREWER'S BLACKBIRD   *Euphagus cyanocephalus*
(Pl. 119)

A gathering of blackbirds sitting on a utility wire near a beach café or walking underneath the cars at a parking lot is not apt to attract attention except, perhaps, from the curious birder. On further scrutiny, the males have iridescent, black plumage and a bright, yellow eye, whereas the females are a plain, dull brown.

The Brewer's Blackbird, formerly a species with a distinctly western range, has rapidly expanded northward and eastward over recent decades, now breeding as far east as the Great Lakes. The success of the Brewer's Blackbird is tied to its adaptability to a variety of human-made habitats. It can be found on lawns, golf courses, and cemeteries from sea level to high elevations. Urbanization poses no problem for the Brewer's Blackbird, for it forages

Plate 119. Brewer's Blackbird: adult male.

wherever it can. Whether picking insects from the front of car grilles in parking lots, or nesting in colonies in street trees along busy thoroughfares, the Brewer's Blackbird is at home almost anywhere.

In winter, Brewer's Blackbirds join with Red-winged Blackbirds *(Agelaius phoeniceus)*, European Starlings *(Sturnus vulgaris)*, and Brown-headed Cowbirds *(Molothrus ater)* to feed and roost in huge flocks. At dusk, a roosting group may congregate by the hundreds at coastal freshwater marshes, their noisy squeaks and whistles issuing from the reeds.

SIZE: Length 9 in. **ADULT MALE:** Iridescent black overall, head tinged with purple. Pale, yellowish eye. **ADULT FEMALE:** Drab gray brown with dark eye. **STATUS:** Common resident. **(B)**

HOODED ORIOLE *Icterus cucullatus*
(Pl. 120)

At a suburban park near the coast or along the promenade of a beach town street, a bright yellow-and-black bird flies out from a palm tree. As it departs, the bird utters an unmusical "weep!" call.

The splendid Hooded Oriole is one of the flashiest of the blackbird tribe. Hooded Orioles nest in palm trees, particularly the California fan palm *(Washingtonia filifera)*. The birds use the long fibers, which they strip from the fronds, to weave hanging nests. The nests are often suspended from the dead fronds that form skirts at the tops of the palm trees.

The Hooded Oriole is a bird of the Southwest, but with the extensive planting of fan palms in residential areas, it has expanded its range as far northward as northern California.

Hooded Orioles feed on insects, supplemented with nectar from blooming exotic plants. By mid-September, they have departed our coast to spend the winter months in Mexico.

SIZE: Length 8 in. **ADULT MALE:** Orange yellow head and underparts, black back and wings with two white wing bars. Black face and throat. **ADULT FEMALE:** Greenish yellow upperparts with two faint wing bars, yellowish underparts. **STATUS:** Fairly common spring migrant and summer resident. **(B)**

Plate 120. Hooded Oriole: adult male (Brad Sillasen).

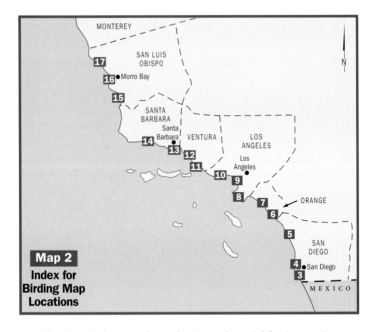

Map 2
Index for Birding Map Locations

The above index map shows the six southern California counties covered by this book. Each number corresponds to a more detailed map showing coastal birding sites. For example, to find birding locations in San Diego County, refer to maps 3, 4, and 5.

SAN DIEGO COUNTY COAST

Perhaps no other destination along the southern California coast is as famous for its birding spots as San Diego County. Routinely included on the birding tour circuit, San Diego has excellent visitor centers and museums, accessible birding spots, and an avifauna that features a mix of southerly and northerly species.

In the last 30 years, the San Diego coastal region has undergone enormous growth. This has resulted in destruction of a great deal of open space; however, the freeways are relatively new and provide access to a number of good birding sites. By carefully studying a map, and avoiding peak rush hours, an observer can take advantage of the San Diego area's abundant bird life.

Southern San Diego County Coast

South San Diego Bay Area (Map 3)

Tijuana Slough National Wildlife Refuge

As the Tijuana River flows to the sea, it creates one of the largest and ecologically richest estuaries in southern California. The Tijuana Slough National Wildlife Refuge and Border Field State Park compose the Tijuana River National Estuarine Research Reserve, which contains approximately 2,500 acres of tidal wetlands and salt marsh. Over 370 species of birds have been documented in the reserve, many of them classified as endangered or threatened. Light-footed Clapper Rails, Western Snowy Plovers, California Least Terns, and Belding's Savannah Sparrows breed at the estuary. Brown Pelicans and an occasional Peregrine Falcon can be spotted, too.

The refuge is located on the Pacific Flyway, so has a variety of wintering waterfowl and migrant and wintering shorebirds. A complete bird checklist is available at the visitor center.

The two best places to bird Tijuana Slough are at the visitor center off Caspian Way and at the end of Seacoast Dr.

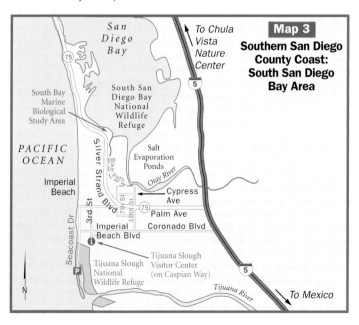

Tijuana Slough Visitor Center and McCoy Trail. The visitor center features regularly scheduled nature walks and programs, educational exhibits, and a good bookstore. Tijuana Slough National Wildlife Refuge, Tijuana Estuary Visitor Center, 301 Caspian Way, Imperial Beach, CA 91932; (619) 575-3613; http://sandiegorefuges.fws.gov/Tijuana.htm.

The visitor center building is surrounded by a native plant garden that lures Say's and Black Phoebes, Common Yellowthroats, White- and Golden-crowned Sparrows (winter), and Western Meadowlarks (winter).

By walking the North McCoy Trail from the visitor center out to the salt marsh, you may glimpse a Light-footed Clapper Rail (early morning best). Herons, ducks, and shorebirds are easily visible in the ponds and channels. Northern Harriers nest in the marsh. This is a good place to look for the occasional Little Blue Heron.

DIRECTIONS (MAP 3): From I-5 in Imperial Beach, take the Coronado Ave. exit (not the Coronado Bridge) west. Follow Coronado Ave., which becomes Imperial Beach Blvd., for approximately 3 miles to 3rd St. Turn left (south) at 3rd St. and then left on Caspian Way into the visitor center parking lot.

End of Seacoast Drive. This portion of the reserve is located in another part of the Tijuana River estuary, near a marsh and within walking distance of the beach. In the marsh, American White Pelicans (uncommon), ducks, and herons forage. To explore the beach, park at the end of Seacoast Dr. and walk over the sand berm near the interpretive signs. Walk left (south) along the beach (watch for nesting Western Snowy Plovers and California Least Terns in spring and summer) to the outlet of the Tijuana River, almost a mile. Elegant Terns roost here in summer and fall. In winter, birds of the open beach such as gulls, Royal Terns, Say's Phoebes, and American Pipits can be seen.

DIRECTIONS (MAP 3): To reach the end of Seacoast Dr. from the visitor center, return to Imperial Beach Blvd. via 3rd St. Turn left (west) on Imperial Beach Blvd., then turn left (south) on Seacoast Dr. and follow it to the end. Parking is limited here and, in summer, can be impossible. Do not leave valuables in your car. In addition to birds, you may spot Border Patrol officers at work.

South San Diego Bay National Wildlife Refuge and South Bay Marine Biological Study Area

In summer, the old saltworks dikes at the south end of San Diego Bay (now designated the South San Diego Bay National Wildlife Refuge) are home to nesting colonies of Double-crested Cormorants, Western Snowy Plovers, American Avocets, Black-necked Stilts, Caspian, Royal, Elegant, Forster's, and California Least Terns, and Black Skimmers, among others. (The rarer Gull-billed Tern *[Sterna nilotica]* also nests here.) A spotting scope is useful because the area is closed to protect the nesting birds; however, while foraging, the cormorants and terns fly right overhead. By late summer, migrant and wintering shorebirds arrive. Depending upon the tides, some of the closer impoundments provide good views of a great variety of shorebirds, including Red Knots, which winter here. Also in winter, this spot attracts hundreds of Brant, plus Lesser Scaup, Redheads, Surf Scoters, and Red-breasted Mergansers. A bike/jogging path runs along the south side of San Diego Bay here. Walk east or west along it.

Another strategy is to drive to the South Bay Marine Biological Study Area (directions below).

Figure 19. California Least Tern.

DIRECTIONS TO THE ENDS OF 7TH AND 10TH STREETS (MAP 3): To reach the ends of 7th St. and 10th St. in Imperial Beach, turn right (north) off Palm Ave. (California 75) onto 10th St., then proceed until 10th St. dead-ends at the edge of San Diego Bay. An even better spot is found by taking Cypress Ave. left (west) off 10th St., then turning right (north) onto 7th St., which also dead-ends bordering the bay.

DIRECTIONS TO THE SOUTH BAY MARINE BIOLOGICAL STUDY AREA (MAP 3): Retrace your route south along 7th St. until it intersects California 75 (Palm Ave.). Turn right on Palm Ave., which becomes Silver Strand Blvd. as it curves north, skirting the edge of San Diego Bay. Approximately a mile along Silver Strand Blvd., look for a large sign saying "Biological Study Area" and a parking pull-out to the right (east). A scope is helpful here.

Chula Vista Nature Center

This impressive Nature Center is surrounded by the Sweetwater Marsh National Wildlife Refuge, a large wetland along the eastern shore of San Diego Bay. Here, where the Sweetwater River meets the bay, an extensive salt marsh is literally at the front door of the interpretive center. A network of trails makes the birding visitor's experience special. Young children will enjoy the live bird enclosures, which include a shorebird and marshbird aviary. General natural history and bird walks are regularly scheduled at the center.

This spot is good for a family outing. For more serious birding, visit in the morning hours so the sun is at your back; midtide is best. The Light-footed Clapper Rail and Belding's Savannah Sparrow are resident species; shorebirds, including Long-billed Curlews, probe the mudflats, and Forster's Terns fly overhead. When the tide comes in, Northern Pintails, Cinnamon Teal, Surf Scoters, and other waterbirds abound. Chula Vista Nature Center, 1000 Gunpowder Point Dr., Chula Vista, CA 91910; (619) 422-2481; www.chulavistanature center.org.

DIRECTIONS (JUST OFF MAP 3): From I-5, exit west at E St. in Chula Vista. Follow E St. west as far as you can, and park your car in the lot (no fee). A free shuttle bus ferries visitors through the locked gate and out the private road to the nature center (fee) in the middle of the marsh.

North San Diego Bay Area (Map 4)

Point Loma Ecological Reserve at Cabrillo National Monument

At Point Loma, which is the peninsula of land that separates North San Diego Bay from the outer coast, the Cabrillo National Monument has a visitor center (fee) featuring early California history and the plants and animals of the San Diego area. The monument includes a Gray Whale *(Eschrichtius robustus)* lookout point, several hiking trails, and panoramic views. Cabrillo National Monument, 1800 Cabrillo Memorial Dr., San Diego, CA 92106; (619) 557-5450; www.nps.gov/cabr.

Just before the entrance kiosk to the monument, take the road to your right (west), which leads down to a shoreline of steep, sandstone cliffs and a couple of parking lots. In winter, scan the rocky shores for Black Oystercatchers (uncommon), Wandering Tattlers, Black Turnstones, and possibly Surfbirds (a scope is helpful). Overhead, a Common Raven usually cruises the thermals. Cormorants (including per-

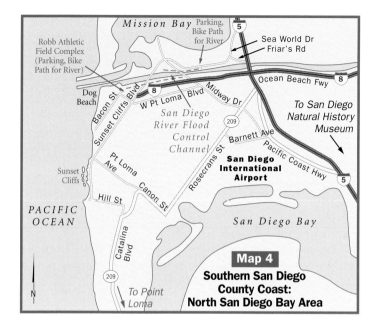

Map 4

Southern San Diego County Coast: North San Diego Bay Area

haps a Pelagic or two) rest on the cliffs, and Black-vented Shearwaters sometimes fly by offshore.

DIRECTIONS (MAP 4): From I-5 south, take California 209 (Rosecrans St.) west; follow Rosecrans St., turn right on Canon St.; turn left (south) onto Catalina Blvd., which becomes Cabrillo Memorial Dr., and follow it to the end. From I-5 north, take Pacific Hwy. to Barnett, then turn left (west) onto California 209.

Sunset Cliffs

Along Sunset Cliffs Blvd. from Hill St. on the south to Point Loma Ave. on the north, stop at pull-outs for good views of the rocky seascape and the birds that frequent this scenic shore. Double-crested, Brandt's—and often a few Pelagic—Cormorants, Ruddy and Black Turnstones, Surfbirds, Wandering Tattlers, and Western and Heermann's Gulls are usually visible in winter. On weekends and in summer, it is crowded with beachgoers.

DIRECTIONS (MAP 4): From Cabrillo National Monument, take Cabrillo Memorial Dr. north; turn left (west) at Hill St.; turn right (north) onto Sunset Cliffs Blvd.

San Diego River Flood Control Channel

One of the top locations for viewing waterbirds is at the San Diego River Flood Control Channel. (Technically, the channel is a part of a large aquatic park, Mission Bay, which has good birding locations nearby to explore on your own.)

Depending upon the light and the tides, the channel features a multitude of species, especially in fall and winter: grebes, herons (including perhaps Little Blue Herons or Reddish Egrets), waterfowl (watch for Eurasian Wigeons) of all kinds including a few Brant, and a variety of shorebirds including Red Knots. On the Robb Field side, look for all of the above plus gulls and terns, and sometimes Snowy Plovers (closer to the river's mouth, near Dog Beach).

DIRECTIONS TO THE NORTH BANK (MAP 4): Take the Sea World Dr. exit off I-5. At the first traffic light on Sea World Dr. beyond Friar's Rd., turn left (south) onto an unmarked road that parallels the river (the road to the north at the intersection is South Shores Park Dr.). Park here and then walk right (west) along the bike path.

DIRECTIONS TO THE SOUTH BANK (MAP 4): Rejoin Sea World Dr. and go west (a left turn at the traffic light when exiting the bike path area). Follow Sea World Dr. as it crosses the river, becoming Sunset Cliffs Blvd. Turn right on West Point Loma Blvd., following it to Bacon St., and turn right (north) into the Robb Athletic Field complex. Drive to the levee path and park; walk to your right (east) or toward the river mouth (west), depending upon which species you seek.

San Diego Natural History Museum

Located in lush Balboa Park, this fine museum is a must for those with an interest in the plants, animals, and geology of San Diego County and Baja California. This is a great place for a family visit. San Diego Natural History Museum, 1788 El Prado, at the east end of Balboa Park; (619) 232-3821; www.sdnhm.org.

DIRECTIONS (JUST OFF MAP 4): From I-5 north or south, exit at Pershing Dr. At the traffic light, turn left onto Florida Dr., proceeding past the naval hospital. Turn left onto Zoo Pl. and go up the hill. Turn left onto Park Blvd., then right onto Village Pl. Turn left into the parking lot behind the museum.

La Jolla Area

La Jolla

The best place in San Diego County for rocky shorebirds and nearshore seabirds is La Jolla. Deep water just off the coast attracts Black-vented Shearwaters from fall through early spring. Wintering loons, cormorants (including Pelagic), scoters, and Royal Terns fly by regularly. The rock ledges and cliffs along the beautiful shoreline attract Wandering Tattlers, Black and Ruddy Turnstones, and Surfbirds from September through April.

DIRECTIONS (NOT ON MAP): Exit I-5 north at Ardath Rd. (Ardath cannot be accessed from I-5 going south; take California 52 west to Ardath Rd.) Follow Ardath Rd. west to La Jolla, where it becomes Torrey Pines Rd.; turn right onto Prospect St. (follow sign to La Jolla Cove). Bear right onto Coast Blvd., which follows the shore. For scanning offshore, the park bordering La Jolla Cove is best. For seeing rocky shorebirds, walk south from the "Children's Pool" area (by the Lifeguard Station where Harbor Seals *[Phoca vitulina]* lounge along the rocks). At the base of Cuvier St., steps lead down to the rocks. Restrooms available. La Jolla's charm attracts tourists, and parking is difficult. Visit on weekdays or very early morning on weekends.

Northern San Diego County Coast (Map 5)

A series of lagoons breach the low-lying shore from La Jolla north to the San Diego County line. Those that have been preserved or restored are good birding spots.

San Elijo Lagoon Ecological Reserve

This lagoon is an example of successful restoration of an estuary, where Escondido Creek meets the coast. At its upper end the lagoon has sizeable freshwater vegetation. Due to periodic dredging, the mouth of the lagoon is kept open, allowing tidal flush over an extensive salt marsh. The marsh and its environs are home to many threatened or endangered species: Light-footed Clapper Rail, Western Snowy Plover, California Least Tern, California Gnatcatcher (rare), and Belding's Savannah Sparrow. Ducks, egrets, shorebirds, Ospreys, Red-tailed Hawks, and Northern Harriers come and go. On the south side, patches of coastal sage scrub hold California Gnatcatchers (rare), Common Yellowthroats, White-crowned and Golden-crowned Sparrows (winter), Spotted and California Towhees, and the Song Sparrows.

The lagoon can be approached on both its north and south sides. Both have good trails. On the north side, a boardwalk through the marsh, a nice, small nature center, and restrooms have been constructed. San Elijo Lagoon Conservancy, P. O. Box 230634, Encinitas, CA 92023-0634; (760) 436-3944 ; www.sanelijo.org.

DIRECTIONS TO THE SOUTH SIDE (MAP 5): At Solana Beach, take Lomas Santa Fe Dr. off I-5. Follow Lomas Santa Fe west to Rios Ave. Turn right (north) on Rios and park at the end of the street (best views of the birds in morning light).

DIRECTIONS TO THE NORTH SIDE (MAP 5): At the next off-ramp north on I-5, take Manchester Ave. west. After about a mile, turn left into the nature center parking lot.

Batiquitos Lagoon Ecological Reserve

A little farther north, Batiquitos Lagoon marks another small watershed, where San Marcos Creek flows into the Pacific Ocean. A level nature trail, which has numbered signposts keyed to an interpretive pamphlet, winds between coastal sage scrub and the salt marsh. Look for Say's Phoebes, Common Yellowthroats, Song Sparrows, and other scrub inhabitants. Watch for wintering grebes, ducks, and shorebirds (a scope is useful). In summer, a fenced-off portion of the lagoon hosts nesting Western Snowy Plovers and California Least Terns. Batiquitos Lagoon Foundation, P.O. Box 130491, Carlsbad, CA 92013; (760) 931-0780, ext. 108: www.batiquitosfoundation.org.

DIRECTIONS (MAP 5): North of Leucadia, exit I-5 at Poinsettia Ln. Travel east on Poinsettia, crossing Paseo del Norte, to Batiquitos Dr. Turn right on Batiquitos, and right again on Gabbiano Way, which ends at a little visitor center and a parking lot. From here, the trail leads east beside the lagoon.

To access the nature trail farther east, return to Batiquitos Dr. and turn right (east) until you come to a "West" and then an "East" parking lot, both of which have paths leading to the lagoon.

Buena Vista Lagoon

Located between Oceanside and Carlsbad, Buena Vista Lagoon is a series of freshwater ponds bordered by cattails and tules. The best part about this site is the nature center and headquarters of the Buena Vista Audubon Society. An attractive building, good interpretive exhibits, and helpful educational materials make this a good family destination. Buena Vista Nature Center, 2202 South Coast Hwy., Oceanside, CA 92054; (760) 439-BIRD; www.bvaudubon.org/center.htm.

The birding, however, would be better if the lagoon were dredged to restore tidal action, a plan being contemplated for the future. Still, the wintering ducks, including Redheads and Canvasbacks, are impressive. Look for Clark's and Western Grebes, Brown Pelicans and American White Pelicans (occasional), Common Moorhens, Soras, Virginia Rails, and Marsh Wrens.

DIRECTIONS (MAP 5): From I-5, exit Vista Way and go west on Vista toward the ocean. Turn left (south) onto Pacific Coast Hwy. (Hwy. S21). Immediately after turning onto Pacific Coast Hwy., begin to watch for the nature center on your left.

ORANGE COUNTY COAST

Orange County, named for the oranges that were once cultivated in acres of orchards, was split from the southeastern portion of Los Angeles County in 1889. In the 1920s, when oil was discovered at Seal Beach, Bolsa Chica, and Huntington Beach, it sparked a wave of development. Today Orange County is California's second most populous county.

The Orange County segment of the southern California coast is not large, but Bolsa Chica Ecological Reserve, Upper Newport Bay Ecological Reserve, and San Joaquin Wildlife Sanctuary are three of the premier birding locations in southern California. Although

Orange County has lost significant wetland resources in the past, recent restoration and management policies have resulted in measurable improvement in the status of several species of wetland birds, including several endangered species.

Southern Orange County Coast (Map 6)

Doheny Beach State Park

This is the best spot for gull-watching in Orange County. In winter, you can find Heermann's, Ring-billed, California, and Western Gulls in large numbers. Among them, study the flock for Bonaparte's, Herring, and Glaucous-winged Gulls.

DIRECTIONS (JUST OFF MAP 6): From Pacific Coast Hwy. (California 1), coming from the south, turn left (west) at Dana Point Harbor Dr. Turn left at the park entrance (fee). Once in the park, drive to the farthest parking lot, from which you can see San Juan Creek flowing into the ocean. Walk around the end of the chain-link fence to get a good view of the gulls.

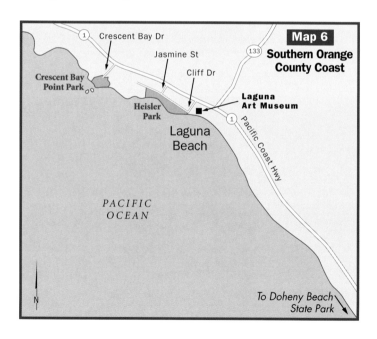

Heisler Park and Crescent Bay Point Park

From Dana Point north to Newport Beach, and especially in the vicinity of Laguna Beach, steep cliffs rise from tidepools and offshore rocks. The scenic shoreline is fringed with expensive houses, but several points afford good views of the rocky coast.

In the midst of the popular art colony and resort of Laguna Beach, Heisler Park provides a grassy strip on the bluffs from which to scan the rocks below for shorebirds. From September through April, look for Black and Ruddy Turnstones, Wandering Tattlers, Surfbirds, and Black Oystercatchers when the rock ledges are exposed at low tide. From Crescent Bay Point Park, more views of the rocky coast can be obtained, including cormorants perched offshore.

DIRECTIONS (MAP 6): In Laguna Beach, turn off Pacific Coast Hwy. (California 1) toward the water on Cliff Dr. (the Laguna Art Museum is on the corner). Cliff Dr. immediately reaches Heisler Park, which has overlooks, picnic benches, and restrooms.

To reach Crescent Bay Point Park, return to Pacific Coast Hwy. and go north, watching for Crescent Bay Dr. on your left. Turn left (west) and drive to the end where a small park overlooks Crescent Bay.

Northern Orange County Coast (Map 7)

Newport Pier

Newport Pier, which extends from the Balboa Peninsula, is perhaps the best place along the Orange County coast to watch for seabirds. In winter and early spring, look for loons, Clark's and Western Grebes, Surf Scoters, and distant Black-vented Shearwaters offshore.

DIRECTIONS (MAP 7): From Pacific Coast Hwy. (California 1), turn (south) onto Newport Blvd., which becomes Balboa Blvd. Follow the signs to the pier. Avoid summer and weekend visits.

Upper Newport Bay Ecological Reserve

Upper Newport Bay is one of the largest estuaries in southern California. With a first-rate interpretive center and many accessible birding sites, this nature preserve deserves top billing.

There are two ways to approach the birding at Upper Newport Bay (often called Newport Back Bay). You can begin at the Peter and Mary Muth Interpretive Center and then bird around the bay, or you can do your birding early, stopping at the center later in the day.

The sparkling new Peter and Mary Muth Interpretive Center is a 10,000-square-foot building constructed almost entirely of renewable or sustainable resources that sits on the northwest end of Upper Newport Bay. The facility is built into the hillside, so it is hidden from the parking lot. In the center, exhibits explain life in the estuary, an excellent video introduces the concept of wetlands, and maps and checklists are available. Because the Interpretive Center is surrounded by Upper Newport Bay Regional Park, there are trails, some of which get close to the bay. Peter and Mary Muth Interpretive Center, 2301 University Dr., Newport Beach, CA 92660: (714) 973-6820; www.ocparks.com or www .newportbay.org.

The birding, however, is better on the opposite (east) side of Upper Newport Bay. A one-way drive, Back Bay Dr., follows along the edge of the bay here for over 3 miles. Birders can stop en route, get out, and observe the ducks, shorebirds, and, with luck, perhaps a Light-footed Clapper Rail. This is the easiest place in southern California to find this secretive species; you will probably hear their "clappering" call. Also look for Soras, Virginia Rails, Marsh Wrens, and Common Yellowthroats. On the brushy hillsides, the coastal sage scrub lures Say's Phoebes, California Towhees, California Gnatcatchers (rare), and other landbirds. Especially in fall and winter, raptors such as Turkey Vultures, Ospreys, White-tailed Kites, Northern Harriers, and Red-shouldered and Red-tailed Hawks can be seen flying overhead.

DIRECTIONS (MAP 7): To get to the Peter and Mary Muth Interpretive Center, which is located near the corner of University Dr. and Irvine Ave., exit Newport Blvd. (California 55) at Del Mar Ave., which becomes University Dr. and dead-ends at the center.

To reach Back Bay Dr., take Jamboree Rd. north off Pacific Coast Hwy. (California 1), and turn left onto Back Bay Dr. (one-way north). A couple of parking lots with interpretive signs are located on Back Bay Dr. The best place is at Big Canyon, where you can park and walk across the street to a freshwater pond.

At its northern end, Back Bay Dr. intersects Eastbluff Dr. There is a viewing platform here, an excellent vantage point for the human-made islands below; a spotting scope is desirable. In winter, gulls and

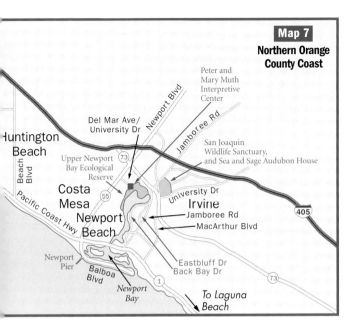

Map 7

Northern Orange County Coast

Peter and Mary Muth Interpretive Center

Del Mar Ave/ University Dr

Huntington Beach

Newport Blvd

Jamboree Rd

San Joaquin Wildlife Sanctuary, and Sea and Sage Audubon House

Upper Newport Bay Ecological Reserve

73

Beach Blvd

Costa Mesa

55

University Dr

Irvine

405

Newport Beach

Pacific Coast Hwy

Jamboree Rd

MacArthur Blvd

Newport Pier

Balboa Blvd

Eastbluff Dr Back Bay Dr

1

73

Newport Bay

To Laguna Beach

shorebirds (including American Avocets and Black-necked Stilts) roost on the islands. In summer, large numbers of Black Skimmers and a few Forster's and California Least Terns nest here.

If you want to walk along Back Bay Dr. rather than drive it, you might decide to park your car along Eastbluff Dr. and walk down the hill west of the viewing platform to Back Bay Dr.

San Joaquin Wildlife Sactuary

Not far inland from Upper Newport Bay in the city of Irvine, a lush freshwater marsh has been restored and opened to the public. Known as the San Joaquin Wildlife Sanctuary, this former private duck club is an excellent birding spot. In the midst of an urban landscape—tall office buildings loom on the horizon—a haven for nature lovers has been created.

The sanctuary, owned by the Irvine Ranch Water District and administered by the San Joaquin Wildlife Sanctuary Board, has developed and contoured the ponds to hold various water levels. The deeper impoundments lure an array of ducks, the shallower ones feature shorebirds. Pied-billed and Eared Grebes, American White

Pelicans (in some years), and White-faced Ibis (uncommon) can be observed. Birds of prey soar overhead: Ospreys, Northern Harriers, and Red-tailed Hawks. Wood Duck *(Aix sponsa)*, Barn Owl, and Tree Swallow nest boxes have been erected. The slopes around the ponds, which are planted with native shrubs and trees, attract migrating and wintering sparrows and warblers.

A network of pathways leads visitors between the ponds, all of which are easily accessible. Many paths are level or have gradual slopes. Benches and restrooms are strategically placed. The sanctuary is an excellent place for elderly or handicapped birders, as well as families with young children.

One of the buildings on the property is the headquarters of Sea and Sage Audubon Society. Staffed by volunteers, the well-stocked Audubon House has a great bookstore, a library, and a variety of educational programs to appeal to the public. Sea and Sage Audubon Society, P.O. Box 5447, Irvine, CA 92612; (949) 261-7963. www.sea andsageaudubon.org.

DIRECTIONS (MAP 7): To reach the San Joaquin Wildlife Sanctuary, exit I-405 at Jamboree Rd. and turn west. At the first traffic light, turn left onto Michelson. Turn right onto Riparian View Dr. and proceed past the Irvine Ranch Water District treatment plant to the signed entrance to the sanctuary.

Bolsa Chica Ecological Reserve

The salt marsh and intertidal mudflats of Bolsa Chica have long been famous for their birdlife. Bolsa Chica was once part of a huge estuary. When oil was discovered, wells and drilling pads peppered the wetlands here. Oil operations will soon be completely phased out, however, as the state of California continues to purchase and restore Bolsa Chica's wetlands in what has been hailed as the greatest example of wetland restoration in southern California.

Future restoration efforts contemplate reconfiguring the coastline by cutting a 360-foot-wide channel through Pacific Coast Hwy. and the adjacent beach, so that a new inlet and tidal basin form. Dredged soil will be turned into levees and nesting islands. Eventually the former oil fields will once again attract the wildlife that have depended upon this estuary for centuries. (Note: At this writing, construction had already begun on the project. Be prepared for changes in the directions below, as traffic and pedestrians along Pacific Coast Hwy. get temporarily rerouted to facilitate construction of the new tidal inlet. The inconvenience should be well worth the result of a much healthier Bolsa Chica wetland.)

An excellent place for birding at any time of year, Bolsa Chica is probably best known for its tern and skimmer colonies in summer.

Caspian, Elegant, Forster's, and California Least Terns nest here, along with a few Royal Terns. Black Skimmers are abundant. The nesting islands are difficult to see without a spotting scope, but the terns fly back and forth overhead frequently.

If you have a spotting scope and want a better view of the nesting islands, it is best to use the southern parking lot (the one just off Pacific Coast Hwy., not the one at the interpretive center) and walk south along the east side of Pacific Coast Hwy. (watch the traffic) about a quarter mile until you come to a sand dune (near the large "Bolsa Chica" sign that faces the other way), where you can look across at the northernmost of the two islands.

The reserve is subject to tidal fluctuations, making the mudflats desirable to a myriad of shorebirds in migration and in winter. Red-necked and Wilson's Phalaropes are found in July and August. When the tide is in, open water lures Clark's and Horned Grebes, as well as the more common Western Grebes. Belding's Savannah Sparrows breed here.

The Bolsa Chica Conservancy, a nonprofit organization, owns and operates the Bolsa Chica Wetlands Interpretive Center. From the interpretive center, you can hike a 1.5-mile trail on a boardwalk with numbered stops keyed to an informative trail map. Bird checklists and other materials are available at the center. Bolsa Chica Wetlands Interpretive Center, 3842 Warner Ave., Huntington Beach, CA 92649; (888) BOLSA4U; www.bolsachica.org.

Then, drive about a mile down Pacific Coast Hwy. to the south parking lot, where a bridge leads over the lagoon and you can watch the birds fly back and forth overhead.

DIRECTIONS (MAP 7): From I-405 or California 22, take the Bolsa Chica Rd. exit south to Warner Ave. Turn right on Warner Ave. and proceed to Pacific Coast Hwy. (California 1). The Bolsa Chica Wetlands Interpretive Center is at the southeast corner of Warner Ave. and Pacific Coast Hwy. To reach the second parking lot, turn left (south) from Warner Ave. onto Pacific Coast Hwy. and go about a mile; watch for the signs on your left (east).

LOS ANGELES COUNTY COAST

Los Angeles County, California's most populous county with its largest city, has a surprising number of good birding locations along its 75-mile coast. With advance planning and a freeway map, you can maximize your bird-watching and enjoy a productive day.

The north coast consists of the steep Santa Monica Mountains, which plunge directly to the sea. Several attractive locations for birding exist here. East of the mountains, the densely urban Los Angeles communities crowd to the coast. From Santa Monica to Redondo Beach, lots of buildings and lots of people make birding difficult.

Southward, the Palos Verdes Peninsula offers striking headlands, steep cliffs, and rocky beaches, which provide excellent birding opportunities, particularly for viewing seabirds (and gray whales) in migration.

Southern Los Angeles County Coast (Map 8)

El Dorado Park and Nature Center

This 85-acre nature sanctuary in Long Beach has native plant areas, nature trails, programs, and classes. Although it is not on the immediate coast, there are two lakes, which have pelicans, cormorants, herons, and ducks in winter. El Dorado Nature Center, 7550 E. Spring

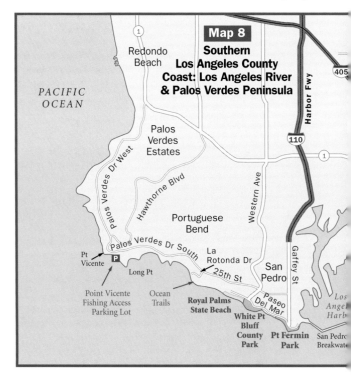

St., Long Beach, CA 90815; (562) 570-1745; www.caohwy.com/e/eldonace.htm.

DIRECTIONS (NOT ON MAP): The park is located off I-605 on Spring St. From southbound I-605, exit at Spring St., go right (west), and the park entrance is on your right. From I-605 northbound, exit at Katella/Willow. Follow Willow to Studebaker Rd., turn north on Studebaker. Go to Spring St. and turn right (east). The nature center is on the south and the park is to the north of Spring St.

Los Angeles River

Between Long Beach on the south and San Pedro on the north, the Los Angeles River empties to the sea. What was once a giant estuary is now an industrial complex comprising the ports of Long Beach and Los Angeles. The birding opportunities are limited. In contrast, a couple of miles upstream, the Los Angeles River, although confined to a concrete channel, can be a productive birding stop. Recent upgrades in landscaping and trails have made the river area more user-friendly.

Time your visit to coincide with low water levels, long after winter rains. In June and July, Black-necked Stilts and American Avocets nest on the concrete abutments and forage in the shallow water. By August and September, thousands of Western Sandpipers and other shorebirds stop here to refuel in migration. Because there are no mudflats for miles north or south, the Los Angeles River provides important feeding and roosting habitat for birds, despite its urban surroundings.

DIRECTIONS (MAP 8): Take I-710 (Long Beach Fwy.) south from I-405. From I-710, exit at Willow St. east. Go over the river and turn left (north) on Golden Ave., then take an immediate left on 26th Way, which dead-ends at the levee. Park on De Forest (watch for streetsweeper schedules). Walk up onto the river levee. A bike path runs on both sides of the river here, and you can walk up- or downstream. The 3-mile stretch between here and Del Amo Blvd. to the north is the best birding.

Palos Verdes Peninsula

Point Fermin Park

This old-fashioned city park has a view of the the harbor, the San Pedro Breakwater, and the rocks below. The historic lighthouse is being refurbished and will soon be ready for tours.

With a spotting scope, in winter and spring scan the rock ledges beneath the park for Black Turnstones, Wandering Tattlers, Surfbirds, and Black Oystercatchers (which nest on the nearby San Pedro Breakwater).

DIRECTIONS (MAP 8): Take I-110 (Harbor Fwy.) south to Gaffey St. in San Pedro. Keep south on Gaffey St. (California 110) until it ends at Paseo del Mar.

White Point Bluff County Park and Royal Palms State Beach

This great spot high above Royal Palms State Beach is perfect for children, because it has play equipment to keep them amused while you scan offshore waters or the rocks below for cormorants, shorebirds, and gulls. Gray Whales can be spotted in February and March. White Point Bluff County Park is free except on weekends. If you see lots of birds, you may wish to drive down the hill to Royal Palms State Beach (fee) for a closer view. Most of the rocky shorebirds can be found here in winter, as well as all three cormorant species.

DIRECTIONS (MAP 8): From the south end of Gaffey St., take Paseo del Mar left (west) until you come to White Point Bluff Park in about 1.5 miles. Turn left into the park (restrooms).

Ocean Trails

This new, 100-acre expanse of coastal sage scrub has terrific potential for birding and whale-watching. Public walkways have been incorporated into a golf course, and the coastal sage scrub plantings boast a variety of bird species. In a few years, the habitat should be even better. Two ponds attract migrant ducks. The seabird-watching is excellent, because of the superior view over the ocean from the bluffs. The southeastern portion is old-growth sage scrub, where California Gnatcatchers (rare), California Towhees, White-crowned Sparrows, Song Sparrows, and other scrub landbirds can be found. The ponds attract Red-winged Blackbirds and Common Yellowthroats.

DIRECTIONS (MAP 8): From White Point Bluff Park, turn left (west) onto Paseo del Mar, turn right on Western Ave. up the hill to 25th St. Turn left onto 25th St., which becomes Palos Verdes Dr. South. Proceed approximately 1.3 miles and turn left (south) on La Rotonda Dr.; follow it down the hill to the Ocean Trails parking lot (restrooms).

Point Vicente Public Fishing Access at Long Point

The best seabird-watching in all of Los Angeles County is from this parking lot just west of a promontory known as Long Point. It is an excellent place to park and set up a spotting scope (which is a necessity here). Restrooms are a bonus. On an April morning, migrating loons, Black-vented and Sooty Shearwaters, terns, and gulls fly by far below. A Peregrine Falcon may be hunting from the cliffs nearby, and Common Ravens swoop about. Hooded Orioles nest in the fan palms to the east of the parking lot. Seabird-watching is best mid-March through early May. The best time for Gray Whale spotting is in March.

DIRECTIONS (MAP 8): From Ocean Trails, travel west on Palos Verdes Dr. South approximately 4.5 miles to Point Vicente Public Fishing Access, a parking lot on the ocean (south) side of the street. You will be unable to turn left from the northbound lane, so go .3 miles to Point Vicente Park and Interpretive Center (now closed), in order to make a U-turn to get back on the south side of Palos Verdes Dr. Then return and enter the Public Fishing Access lot.

Central Los Angeles County Coast (Map 9)

Natural History Museum of Los Angeles County

Although not located on the coast, this wonderful museum — third largest of its kind in the United States — is mentioned because it has a wealth of information on the area's natural history. The museum is located in Exposition Park, across from the University of Southern California (USC), in downtown Los Angeles. Natural History Museum of Los Angeles County, 900 Exposition Blvd., Los Angeles, CA 90007; (213) 763-DINO; www.nhm.org.

DIRECTIONS (NOT ON MAP): Take Exposition Blvd. west off I-110 (Harbor Fwy.). Parking is available in patrolled lots off Menlo Ave. Enter from Vermont.

Playa del Rey Jetties

One of the good bird-finding spots between Malibu and the Palos Verdes Peninsula is the rock jetties at the entrance to the Marina del Rey harbor. In addition, upstream along Ballona Creek, much of the Ballona wetlands complex is slated for restoration and should be increasingly productive for birding in the future.

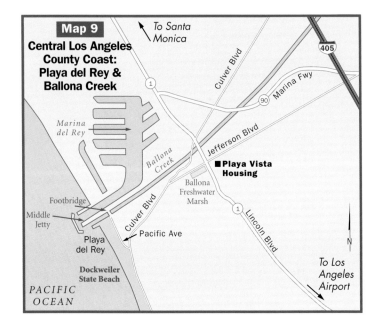

Map 9

Central Los Angeles County Coast: Playa del Rey & Ballona Creek

To Santa Monica

405

Culver Blvd

Marina Fwy

90

Marina del Rey

Jefferson Blvd

Ballona Creek

■ Playa Vista Housing

Ballona Freshwater Marsh

Footbridge

Middle Jetty

Culver Blvd

1

Lincoln Blvd

Pacific Ave

Playa del Rey

N

Dockweiler State Beach

To Los Angeles Airport

PACIFIC OCEAN

The middle jetty is a level walkway that affords close-up views of rocky shorebirds such as Surfbirds, Ruddy and Black Turnstones, Wandering Tattlers, and maybe a Black Oystercatcher. Other shorebirds come and go. Cormorants, pelicans, gulls, and terns roost nearby, some of them on neighboring Dockweiler State Beach. In winter, scoters, Red-breasted Mergansers, and Eared, Western, and Clark's Grebes swim in the channels.

DIRECTIONS (MAP 9): From I-405, take California 90 (Marina Fwy.) west to Culver Blvd. Turn left (south) on Culver. At the end of Culver (do not bear left at the turn), turn right onto Pacific Ave. and go several blocks to its end. Find a parking place as close as you can (it is crowded on weekends) and walk across the footbridge at the end of Pacific Ave. to the middle jetty. Walk left out on the jetty, or right following the path northeast along the Ballona Creek channel.

Ballona Freshwater Marsh

This recently restored (2003) wetland has cattails and bulrushes lining a trail around a lagoon. Many freshwater marsh species such as White-faced Ibis (fall), Virginia Rail, and Red-winged Blackbird can be found. Look for Red-necked and Wilson's Phalaropes, as well as dabbling ducks in season.

DIRECTIONS (MAP 9): Park along the south side of Jefferson Blvd. and enter through three places in the chain-link fence.

Northern Los Angeles County Coast (Maps 10a, 10b)

Malibu Lagoon State Beach

Right beside busy Pacific Coast Hwy. (California 1), Malibu Lagoon always turns up something interesting for birders. At almost any season, the lagoon, which is usually enclosed by a sandbar, shelters an assortment of grebes, pelicans, herons, terns, and shorebirds. Numerous ducks winter here. In the marshy areas, you may glimpse a Sora or Virginia Rail. Swallows course overhead, Black Phoebes patrol the shores, and gulls of various kinds loaf on the sand. The Santa Monica Bay Audubon Society welcomes families and beginners for monthly

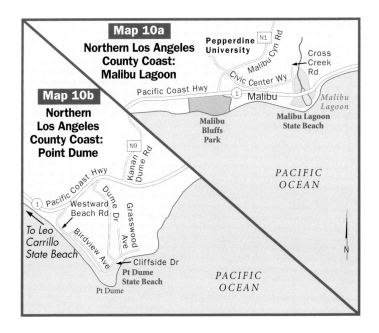

bird walks at the lagoon. Santa Monica Bay Audubon Society, P.O. Box 35, Pacific Palisades, CA 90272.

DIRECTIONS (MAP 10A): Malibu Lagoon is part of Malibu Lagoon State Beach. The entrance is on the south side of the intersection of Cross Creek Rd. and PaciWc Coast Hwy., just west of where the highway crosses Malibu Creek. You may choose to park in the parking lot of the state beach (fee), or you can park along the south side of Pacific Coast Hwy. and walk in. (Be sure not to leave valuables in your car.) Once inside the park entrance, take the trail out to your left (east), stopping at good vantage points to view the lagoon. Restrooms.

Point Dume State Beach

On a spring day when the yellow-flowering giant coreopsis *(Coreopsis gigantea)* are blooming and the Pacific Ocean sparkles below, the view alone is worth a visit. This spectacular headland is used by Los Angeles area birders for watching seabirds in migration. In early morning during March and April, a succession of migrating loons, cormorants, brant, scoters, gulls, and terns fly by on their way north. A boardwalk leads through the restored sand dune vegetation to a comfortable wooden viewing platform with seats. A spotting scope is helpful.

DIRECTIONS (MAP 10B): Approximately 6 miles west of Malibu Lagoon, Point Dume can be reached by turning left (south) off Pacific Coast Hwy. by two routes: (1) Take Dume Dr. to Cliffside Dr., turn right (west), and you'll see eight parking spaces and the county beach sign; or (2) take Westward Beach Rd., which becomes Birdview Ave. and ends at the parking spaces for Point Dume. (If parking is full, you may have to go east on Cliffside Ave. to Grasswood Ave., where street parking is available.) No restrooms.

Leo Carrillo State Beach

From September through April (early morning best), this is one of the prettiest and least crowded of the state beaches. The cove has a sandy beach with rock stacks offshore. Shorebirds of various kinds stop here in spring and fall migration. There are always plenty of gulls to study in winter. Scoping for seabirds from the bluffs near the lifeguard station can yield shearwaters.

DIRECTIONS (JUST OFF MAP 10B): Approximately 8 miles west on Pacific Coast Hwy., watch for the Leo Carrillo State Beach sign. Go past the beach sign, make a U-turn (with care), and return to where you can legally park (watch the signs) along the south side of the highway east of the little creek. Walk back to where the bridge goes over the creek, on the east side of which is a pathway to the beach. Restrooms.

VENTURA COUNTY COAST

Ventura County is a breath of fresh air to birders escaping the hectic pace of the freeways to the south. Most of the birding sites are easy to find and uncrowded.

Ventura County's 43 miles of coastline hug the Santa Monica Mountains to the south and the Santa Ynez Mountains to the north. In between, the Santa Clara River traverses the Oxnard Plain to the sea. The cities of Ventura, Oxnard, and Port Hueneme are located on this wide, fertile floodplain. Once a largely agricultural region, the county has seen its citrus orchards and row crops rapidly overtaken by housing developments.

The richest area for birds is the impressive Santa Clara River estuary, which encompasses a number of fascinating wetland habitats.

The Channel Islands National Park Visitor Center, located at Ventura Harbor, has a wealth of information about the Channel Islands and how to get to them.

Southern Ventura County Coast (Map 11)

Mugu Rock Overlook

Mugu Rock, located at Point Mugu, is part of Point Mugu State Park. The rock is an enormous dome that looms on the ocean side of the Pacific Coast Hwy. There are places to pull out and set up a spotting scope on either side of the rock. The southeast side is most rewarding, depending upon the time of day. In spring and fall, scan for migrating loons, scoters, and Sooty Shearwaters as they fly offshore. In winter, Black-vented Shearwaters are possible. Gray Whales migrate along the coast in February and March. Bring a scope if possible.

DIRECTIONS (MAP 11): Mugu Rock can be reached by taking Las Posas Rd. south off U.S. 101 to Pacific Coast Hwy. (California 1). The rock is approximately 2 miles farther south on the right (west) side of the highway. If you approach from the south, you will have to go beyond Mugu Rock, until you come to a place where you can make a U-turn to return to the rock. Do not cross the double yellow line to do so.

Mugu Lagoon

Mugu Lagoon is an extensive salt marsh located within Point Mugu Naval Air Station. As such, it is off-limits to the public; however, a wide dirt pull-out with an interpretive sign and a viewing area for the tidal marsh is accessible off Pacific Coast Hwy. From here, herons and

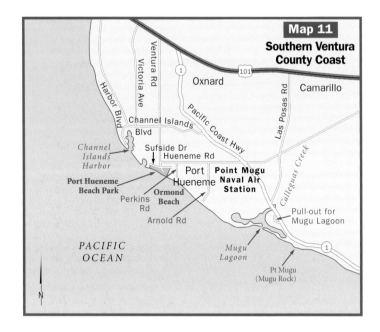

Map 11
Southern Ventura County Coast

egrets, shorebirds, and ducks can be seen in fall and winter, although many of the birds are far away. The easternmost mudflats are often the most productive.

DIRECTIONS (MAP 11): To reach Mugu Lagoon, follow directions above for Mugu Rock. Just after turning from Las Posas Rd. onto Pacific Coast Hwy. (California 1), watch for the place where the road ceases to be a freeway. Then, drive into the wide dirt pull-out located on your right (west). A chain-link fence separates you from the marsh. The directions above for approaching Mugu Rock from the south apply to Mugu Lagoon as well.

Ormond Beach

The Ormond Beach area is the future site of an ambitious wetlands restoration project. The state Coastal Conservancy is committed to preserving the salt marsh behind Ormond Beach and the surrounding acreage, forming a link in the wetlands complex that stretches nearly 9 miles from Point Mugu to Port Hueneme. Given that the newly acquired area is bordered by several industrial operations, this is somewhat difficult to imagine. The new project will make the whole Ormond Beach complex an even more rewarding birding destination in the future.

The beach itself is a long expanse of flat sand backed by dunes and salt marsh. In spring and summer, California Least Terns and Western Snowy Plovers nest here in good numbers. The wetlands trail at the end of Perkins Rd. offers good looks at herons and egrets, common shorebirds in migration, and ducks in winter.

DIRECTIONS (MAP 11): There are two ways to get to Ormond Beach: one from Port Hueneme Beach Park, the other at the end of Arnold Rd. In between is a loop trail around the wetlands, but it is impossible to access the outer beach from the wetlands.

From Port Hueneme Beach Park: Take Victoria Ave. south from U.S. 101 (Ventura Fwy.). Turn left (east) on Channel Islands Blvd. Turn right (south) onto Ventura Rd. and follow it to its end. Take a left (east) at Surfside Dr., turn right on Ocean View, and park in the lot at the end (fee). To reach the tern and plover nesting colonies, walk left (east) along the beach for about 2 miles.

From Arnold Rd.: To reach the southeast end of the tern colony, park at the end of Arnold Rd., which is off Hueneme Rd., and walk northwest.

To the wetlands at the end of Perkins Rd.: From Ventura Rd. turn left (east) onto Hueneme Rd. After about a half mile, turn right (south) on Perkins Rd. and follow it to the end to a free parking lot. A small footpath and footbridges make it easy to walk around the wetlands.

Channel Islands Harbor

In spring, a colony of Great Blue Herons and Black-crowned Night-Herons nests in the cypress trees adjacent to the boat slips. Hooded Orioles build their hanging nests in the fronds of the California fan palms *(Washingtonia filifera)* nearby. The whole harbor area is slated for a giant remodel, which may threaten the heron colony.

DIRECTIONS (MAP 11): From Channel Islands Blvd., turn left (south) onto Harbor Blvd. Follow it to Barracuda Way, turn left, and park.

Northern Ventura County Coast (Map 12)

The Santa Clara River, draining one of the largest watersheds in southern California, meets the sea in a mosaic of habitats. A nice assortment of birds can be found in any season. McGrath State Beach, Ventura Water Treatment Plant/Wildlife Ponds, Surfer's Knoll off Spinnaker Dr., and Ventura Harbor are all part of the estuary complex.

Because the mouth of the river is usually blocked by a sandbar, a large lagoon forms in years when rainfall is low and runoff is negligible. Under these conditions grebes, cormorants, pelicans, herons, and ducks use the lagoon. When the sandbar is breached, either by high tides or winter storms, tidal mudflats are exposed, creating perfect conditions for shorebird feeding. Many, many species of shorebirds have been spotted at the estuary, best in late summer and fall. In fall and winter, a Peregrine Falcon or a Merlin may be seen. Below are some suggestions on how to access the Santa Clara River estuary.

McGrath State Beach

From the campground, situated south of the estuary mouth, you can walk out to the beach or explore a nature trail through the willows to the edge of the river. At the beach, a variety of terns, including Elegant, Forster's, and California Least are present in summer. Western Snowy Plovers nest nearby. Alternatively, by taking the nature trail, you reach the edge of the river channel farther upstream. Remember to bring wading boots if you want to walk out onto the mudflats; in recent years, water levels have regularly been too deep to be able to do so. Even if you cannot walk out in the mud, a variety of waterbirds is usually close enough for good views, although a spotting scope is helpful.

DIRECTIONS (MAP 12): In Ventura, take Seaward Ave. west from U.S. 101. Turn left (south) on Harbor Blvd. Stay on Harbor Blvd., going past Spinnaker Dr. and continuing across the Santa Clara River on a bridge. Immediately after the bridge, turn right (west) into McGrath State Beach (fee). Park at the day use area (at the far right or north end of the campground) and walk straight ahead (west) on a trail to the beach, or north on a trail through the willows.

Ventura Harbor and Channel Islands National Park Visitor Center

Ventura Harbor, located north of the Santa Clara River estuary, is important to birders because it is the headquarters of Channel Islands National Park. In addition, it has good opportunities for finding loons, grebes, and diving ducks in the harbor waters around the visitor center and the nearby boat docks. Across from the visitor center in fall through spring, rocky shorebirds such as Wandering Tattlers, Surfbirds, and Black Oystercatchers forage on the two jetties that protect the harbor's entrance.

At the visitor center, information is available on how to reach the Channel Islands by boat or by plane. Even if you are not planning to visit the islands, take time to learn about them here. An excellent bookstore has maps, checklists, a film, and exhibits about features

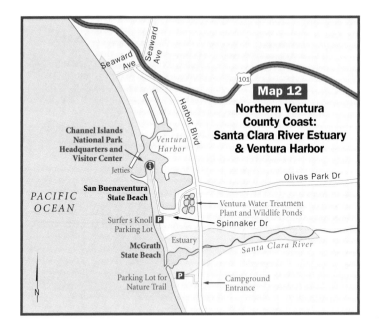

Map 12

**Northern Ventura
County Coast:
Santa Clara River Estuary
& Ventura Harbor**

of the islands. Outside the building, a native plant garden with numbered signs is keyed to an informative pamphlet. Channel Islands National Park, 1901 Spinnaker Dr., Ventura, CA 93001; (805) 658-5730; www.nps.gov/chis.

DIRECTIONS (MAP 12): Turn west onto Spinnaker Dr. from Harbor Blvd. Follow it to the end and park at the visitor center.

Surfer's Knoll

From this spot on the beach, a relatively short walk (half a mile) puts you at the northern edge of the Santa Clara River mouth. In summer, nesting California Least Terns sit on eggs as you walk by the fence that protects them. Sometimes Western Snowy Plovers share this site for breeding, too. Nearing the river mouth, walk slowly, as there may be nesting American Avocets and Black-necked Stilts among the sparse vegetation on the sand dunes. The muddy fringes of the estuary waters here usually support a variety of shorebirds, especially August through October. Of course, if the estuary has drained, it will be packed with shorebirds. If not, the birding is still great. Something exciting is always going on here: shorebirds, gulls, terns, and ducks abound. In fall and winter, look for a Peregrine Falcon or Merlin.

DIRECTIONS (MAP 12): From Harbor Blvd., follow Spinnaker Dr. and turn left into the parking lot (restrooms) at the point where Spinnaker makes a sharp-angle turn to the right. If this lot is full, park across the street at Ventura Harbor Village. Walk southeast along the beach past the tern colony fence to the Santa Clara River mouth.

Ventura Water Treatment Plant/Wildlife Ponds

The ponds here attract waterfowl, marshbirds, and swallows. By walking along the dikes between the ponds in fall and winter, you can see grebes, herons and egrets, ducks, Soras, Common Moorhens, phalaropes, Bonaparte's Gulls, Marsh Wrens, and Common Yellowthroats. Swallows are abundant in spring migration, and Tree Swallows use the nest boxes in summer.

DIRECTIONS (MAP 12): The Ventura Water Treatment Plant is located at 1400 Spinnaker Dr., a right turn if you are on Spinnaker headed toward Harbor Blvd. from Surfer's Knoll. Once in the plant, follow the signs for the administration building, where you must sign in before you proceed. Currently, the plant is closed on weekends.

SANTA BARBARA COUNTY COAST

Santa Barbara County's coastline is unique. The northern coast, from the cliffs at Point Conception to the Santa Maria River mouth, is characterized by cooler nearshore waters, rocky cliffs, and isolated beaches. The 35 miles of undeveloped coastline on Vandenberg Air Force Base is inaccessible to the public; as such, it protects sizeable populations of breeding Western Snowy Plovers and California Least Terns as well as many other bird species.

Santa Barbara's southern coast, from Gaviota to Carpinteria, has calmer, warmer offshore waters and a gentler climate. Much of the shoreline is marked by urban development, but an enlightened populace and a strong conservation ethic have helped preserve or restore a surprising amount of habitat for birds. Indeed, coastal Santa Barbara County is legendary for its wealth of birdlife and active birding community. The locations listed below are a few of the best.

Southern Santa Barbara County Coast

Carpinteria Salt Marsh Nature Park

This small wetland shows what restoration efforts can do. Originally part of the larger Carpinteria Salt Marsh administered by the University of California Reserve system, this remnant is now jointly maintained with the city of Carpinteria. Where once there was a dry wasteland of trash and weeds, tidal channels have been renewed and native plants grace the pathways. A mile-long walk tours the wetland. Herons, egrets, shorebirds, Merlins (uncommon), White-tailed Kites, Northern Harriers, Ospreys, Belted Kingfishers, and Belding's Savannah Sparrows can be seen, depending on the season.

Docent tours are given at 10:00 A.M. on Saturday mornings; for further information contact the city of Carpinteria at (805) 684-5405.

DIRECTIONS (JUST OFF MAP 13): Take the Casitas Pass Rd. exit off U.S. 101, turn south to Carpinteria Ave. Turn right (west) onto Carpinteria Ave. and left (south) onto Linden Ave. Follow Linden Ave., and just before the street ends, turn right (west) on Sandyland Rd. Follow Sandyland Rd. to Ash Ave., turn right on Ash Ave., and park. There are interpretive signs and restrooms.

Santa Barbara Waterfront Area (Map 13)

Andree Clark Bird Refuge

This brackish lake lined with reeds attracts a surprising array of waterfowl and gulls in fall and winter. Glaucous-winged, Western, California, Ring-billed, and Heermann's Gulls sometimes rest and preen near the parking lot. Black-crowned Night-Herons and Green Herons roost on the island vegetation in the middle of the lake. Marsh Wrens (winter) and Common Yellowthroats abound. In spring, Double-crested Cormorants nest in the eucalyptus trees at the far west end of the refuge.

DIRECTIONS (MAP 13): Exit U.S. 101 at Hot Springs Rd. Take Cabrillo Blvd. south, then turn right at Los Patos Way. The lake and parking lot are on the left (west). Follow the trail along the north side of the lake to reach several good viewing platforms.

East Beach

Two spots on East Beach—at the base of Garden St., where an outfall channel often creates a large pool in the sand, and at the nearby mouth of Mission Creek—lure gulls (including Mew and Glaucous-winged) and terns from early fall through spring. Elegant Terns are fairly common in summer and early fall. In winter, a large flock of Black Skimmers roosts on East Beach; the group may be anywhere in either direction along the sand. Common shorebirds can be seen, too, especially in spring and fall.

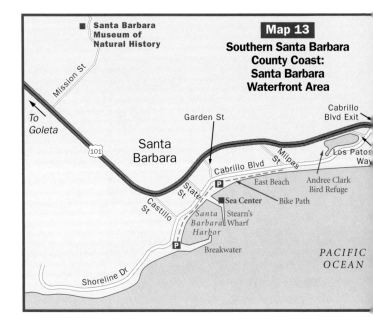

Map 13

Southern Santa Barbara County Coast: Santa Barbara Waterfront Area

Santa Barbara Museum of Natural History

Mission St

To Goleta

101

Santa Barbara

Garden St

Cabrillo Blvd Exit

Los Patos Way

Milpas St

Cabrillo Blvd

State St

Castillo St

East Beach

Andree Clark Bird Refuge

■ Sea Center

Santa Barbara Harbor

Stearn's Wharf

Bike Path

P

P

Breakwater

Shoreline Dr

PACIFIC OCEAN

DIRECTIONS (MAP 13): Exit U.S. 101 south at Garden St. and follow Garden St. until it ends at a parking lot (fee on weekends and in summer) south of Cabrillo Blvd. (Alternatively, drive from the bird refuge along Cabrillo Blvd. west until it intersects Garden St., then turn left (south) into the parking lot by the beach.) From the parking lot, walk east along the bike path to the outfall, or west to the mouth of Mission Creek, or walk out to Stearn's Wharf beyond. On weekdays, parking is usually available along Cabrillo Blvd. Avoid weekends in summer, except early in the morning.

Santa Barbara Harbor

The harbor shelters wintering loons, grebes, and diving ducks, as well as gulls, terns, and some shorebirds. From fall through spring, a walk out to the end of the breakwater should yield rocky shore bird species such as Black and Ruddy Turnstones. Wandering Tattlers and Surfbirds appear in spring and fall migration. Time your visit at middle or low tide, so the rocks will be exposed.

DIRECTIONS (MAP 13): To reach the harbor area from the parking lot at the base of Garden St., turn left (west) onto Cabrillo Blvd. and proceed approximately a mile to the harbor entrance, which is be-

To Carpinteria, Salt Marsh Nature Park →

Hot Springs Rd

101

N

yond the traffic light at Castillo St. Turn left (south) into the harbor entrance. Parking lots are available to the right and to the left (fee). To reach the breakwater, follow the sidewalk past the restaurants and a maritime museum.

Santa Barbara Museum of Natural History and the Sea Center

The Santa Barbara Museum of Natural History is an exceptional institution, with a main campus located in Mission Canyon and a satellite facility, the Ty Warner Sea Center, on Stearn's Wharf. The Sea Center, which emphasizes marine natural history, is recently refurbished with state-of-the-art exhibits and has just reopened to the public— an excellent stop for families. Ty Warner Sea Center, 211 Stearns Wharf, Santa Barbara, CA 93101; (805) 962- 2526; www.sbnature.org/ seacenter/.

The main museum has a number of excellent exhibits about birds of the region, plus a gift store where a good selection of natural history books and checklists is on display. Santa Barbara Museum of Natural History, 2559 Puesta del Sol Rd., Santa Barbara, CA 93105; (805) 682-4711; www.sbnature.org.

DIRECTIONS (MAP 13): To visit the Sea Center, walk or drive onto Stearn's Wharf, which is at the base of State St.

To visit the Museum, exit U.S. 101 east at Mission St. Follow signs to the Old Mission and the museum, which is located directly behind the Old Mission off Mission Canyon Rd.

Goleta Area (Map 14)

More Mesa

More Mesa is a remnant of coastal grassland mixed with coastal sage scrub on the bluffs overlooking the ocean. One of the few parcels of untouched open space on the south coast, it is accessible to the public by a network of unimproved trails. In late fall and winter, you may see White-tailed Kites, Northern Harriers, Barn Owls, Short-eared Owls *(Asio flammeus)* (rare), Say's Phoebes, Savannah Sparrows, and Western Meadowlarks. A visit at dusk is best for the owls.

DIRECTIONS (MAP 14): To reach More Mesa, exit U.S. 101 at S. Patterson Ave., turn left (south), and proceed, crossing Hollister Ave. When the road (which becomes Shoreline Dr.) takes a sharp left turn, watch for a parking pull-out on the opposite side of the road (past the stables). Park there and walk south along Shoreline Dr. up a hill. Take the first trail on your left that heads east onto the mesa (before the houses).

Goleta Beach County Park

Goleta Slough drains into a tidal lagoon at Goleta Beach. Wintering loons, grebes, diving ducks, gulls, and terns frequent the channel north of the parking lot across from the restaurant. At low tide, look for common shorebirds. A Belted Kingfisher or two are often nearby. Sometimes a Peregrine Falcon perches on trees atop the nearby bluffs. From late winter through spring, a colony of Great Blue Herons nests in the eucalyptus grove on the far side of the tidal channel. For closer looks at grebes, loons, cormorants, and scoters, walk out on the pier; watch for Barn Swallows nesting underneath it in spring.

DIRECTIONS (MAP 14): From U.S. 101 (sign to UCSB), take California 217 (Clarence Ward Memorial Blvd.) south and then exit at Sandspit Rd. Turn left (south) at the intersection and follow the signs to Goleta Beach. After entering the park, turn into the left-hand (eastern) parking areas.

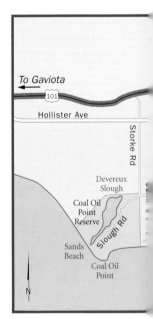

Goleta Point at UCSB

The best spot along the southern Santa Barbara coast for watching seabirds is at Goleta Point (Campus Point), reached from the University of California at Santa Barbara campus. This lookout from a spot on the bluffs is an excellent place in spring for watching the northward seabird migration. A scope is desirable. Large flocks of loons (mostly Pacific and Red-throated), Brant, scoters, gulls, and terns fly up the coast close to shore. Best viewing time is late March to mid-May. Also at this season, Surfbirds, Ruddy and Black Turnstones, and Wandering Tattlers pause in spring migration to feed on the rocks below the bluffs. Offshore, look for Sooty Shearwaters from spring through fall, and Black-vented Shearwaters in winter. Also in winter, on the campus lagoon behind Goleta Point, Redheads, Greater *(Aythya marila)* (rare) and Lesser Scaup, and Red-breasted Mergansers may be found.

DIRECTIONS (MAP 14): To reach Goleta Point, exit U.S. 101 at California 217 (Clarence Ward Memorial Blvd.), just as you would for Goleta Beach County Park. Instead of turning off at Sandspit Rd., remain on California 217 until it ends at the entrance kiosk for UCSB, where information about parking may be available. Parking regulations at the UCSB campus are subject to change. Currently, visitor lots have been designated, where a self-pay permit system operates. For more information see www.tps.ucsb.edu. After the entrance kiosk, turn left (south) on Lagoon Rd. until you come to a parking lot on your left (east) where the Marine Science Institute building is located (metered parking usually available). Walk west past the Marine Science Building and down to the beach, then up a trail to the bluffs; bear left to where an old piece of concrete foundation marks the overlook spot.

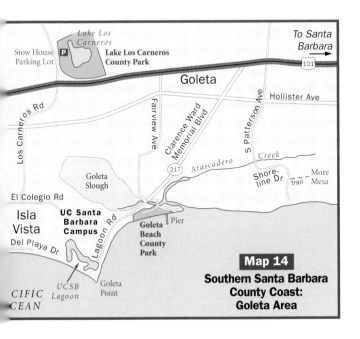

Map 14

Southern Santa Barbara County Coast: Goleta Area

Coal Oil Point Reserve/Devereux Slough

Devereux Slough, part of Coal Oil Point Reserve managed by the University of California at Santa Barbara, is one of the rare examples of salt marsh habitat remaining on Santa Barbara's south coast. A network of trails leads around the slough.

In winter, egrets, herons, a variety of waterfowl, gulls (including Mew), and many shorebirds visit the slough, depending upon water levels. Occasionally, the slough may dry up for a good portion of the fall, in which case fewer birds are found. Belding's Savannah Sparrows nest every spring in the pickleweed bordering the marsh. Red-tailed Hawks and White-tailed Kites circle overhead, and swallows pass through in spring migration.

Out at Sands Beach, where a sandbar blocks the mouth of Devereux Slough, a large colony of Snowy Plovers plus other interesting shorebirds spend the fall and winter. Some of the Snowy Plovers have recently begun to breed here in summer. For more information about Coal Oil Point Reserve see www.Coaloilpoint.ucnrs.org.

DIRECTIONS (MAP 14): Parking regulations on this West Campus of UCSB are strictly enforced; however, Coal Oil Point Reserve managers encourage visiting birders. Here are some suggestions: Exit U.S. 101 south at Storke Rd./Glen Annie Rd. off ramp. Take Storke Rd. south. When Storke intersects with El Colegio Rd. at a sharp left angle, bear straight ahead onto Slough Rd.

To park at the Cliff House parking lot at the end of Slough Rd., you need a permit (fee) at all times. Contact the reserve's director in advance at (805) 893-5092 or sandoval@lifesci.ucsb.edu.

Figure 20.
Western
Snowy Plover.

At any time, birders may drive along Slough Rd. and park in several pull-outs to observe the slough. Currently, as long as you stay within view of your car, you will not be ticketed.

If you want to park and walk on the trails or on the beach (to see the plover colony) at the reserve and are not able to get a permit in advance, park at the far west end of Del Playa Dr. in Isla Vista and walk west along the bluffs to Coal Oil Point. Sands Beach is west of Coal Oil Point.

Lake Los Carneros County Park

A trail and a footbridge make it easy to access this freshwater lake, and the birding is always pleasant. Although not strictly coastal, Lake Los Carneros was once part of the extensive wetlands complex that composed Goleta Slough and is included here as an example of freshwater marsh habitat. A walk around the lake from September through April should yield Soras, Virginia Rails, Common Moorhens (uncommon), and Marsh Wrens. Red-winged Blackbirds sing from the reedbeds. In winter, a variety of ducks float on the lake, and birds of prey (occasionally, a Merlin) soar overhead or perch nearby. Flocks of Golden-crowned and White-crowned Sparrows along with California Towhees inhabit the patches of coastal sage scrub around the lake in fall and winter.

DIRECTIONS (MAP 14): Exit U.S. 101 at Los Carneros Rd. and proceed north. Watch for a fire station on your right (east), in front of which a road leads to the parking lot for Stow House. Lake Los Carneros is east of the house.

Northern Santa Barbara County Coast (Map 15)

Guadalupe Dunes Preserve

The Santa Maria River flows across the agricultural plain west of the city of Santa Maria, emptying into the Pacific Ocean at Rancho Guadalupe Dunes County Park (now part of the Guadalupe-Nipomo Dunes Preserve). To learn more about the coastal dune ecosystem here, stop at the Dunes Center (see below).

In fall and winter, the beach at the river mouth is a good place to scan offshore for birds such as loons, grebes, and scoters. Sooty Shearwaters are seen in summer or early fall, and Black-vented Shearwaters in late fall and early winter.

In spring and summer, nesting Western Snowy Plovers and California Least Terns occupy the dunes on both sides of the river mouth.

For this reason, access to the river estuary itself is prohibited from March 1 through September 30, to avoid potential disturbance by humans. Outside those dates, a walk north of the parking lot to view the river mouth is rewarding, depending upon water levels. Birding is best for shorebirds when the water is low; however, pelicans, gulls, and terns are always abundant, and Peregrine Falcons often hunt here, too.

DIRECTIONS (MAP 15): Exit U.S. 101 at Main St. (California 166) in Santa Maria and proceed west approximately 11 miles to the kiosk at the entrance to the preserve, and then continue on toward the beach parking lot (restrooms).

Dunes Center

This small museum, occupying a 1910-era house on Guadalupe's main street, is fun, especially for families with children. Through education and research, the Dunes Center promotes conservation of the Guadalupe-Nipomo dunes ecosystem, the second largest stretch of dunes in California. A good selection of checklists and books is available about the birds and vegetation of the region. The Dunes

Center, 1055 Guadalupe St. (California 1), Guadalupe, CA 93434; (805) 343-2455; www.dunescenter.org.

DIRECTIONS (MAP 15): The Dunes Center is located on California 1 (Guadalupe St.) 2 miles north of its intersection with California 166, in the town of Guadalupe.

SAN LUIS OBISPO COUNTY COAST

The pristine coastline of San Luis Obispo County, featuring rugged cliffs, peaceful coves, and grassy terraces, offers exciting bird-watching opportunities. The farther north you travel, the more the birdlife resembles that of northern California.

Midway along the coast, Morro Bay stands out as an important sanctuary for many kinds of birds. You seldom encounter a whole community that considers itself a wildlife sanctuary, but Morro Bay deserves the honor. South of Morro Bay, at Oso Flaco Lake and Shell Beach, more good birding can be found.

Pick your target species, and then design a memorable birding trip along the coast of this beautiful county.

Southern San Luis Obispo County Coast (Map 15)

Oso Flaco Lake

The northern portion of the Guadalupe-Nipomo Dunes Preserve encompasses Oso Flaco Lake, an unusual freshwater lake nestled in the midst of sand dunes. The birding here is good in winter and spring. In winter, all sorts of ducks swim on the lake, American Bitterns *(Botaurus lentiginosus)* (uncommon) hide in the reeds, and a Peregrine Falcon might swoop down any time. Also look for American White Pelicans, Common Moorhens, Virginia Rails, and Soras. In spring, terns (Caspian and Forster's) and swallows course above the water. Barn Swallows nest under the boardwalk, Marsh Wrens nest in the tules, and Tree Swallows nest in the cavities of the willow trunks. A California Least Tern colony is out at the beach. A convenient boardwalk allows you to walk across the lake and out among the sand dunes to the Pacific Ocean.

DIRECTIONS (MAP 15): To get to Oso Flaco Lake, follow California 1 north of Guadalupe approximately 4.5 miles until you come to Oso Flaco Lake Rd. Turn left (west) and continue to the parking lot (fee) at the end (restrooms). From Nipomo, take Tefft St. west off U.S. 101; turn left on Orchard Rd., turn right on Division St., and right onto Oso Flaco Lake Rd.

Cliffs at Shell Beach and Margo Dodd Park

In spring, when the Pelagic Cormorants and Pigeon Guillemots are nesting on the cliffs here, good views can be had from overlooks on adjoining bluff tops. A pair of Peregrine Falcons usually breeds nearby, and the young make practice flights at eye level. Below, Black Oystercatchers set up territories on the rocks. Western Gulls nest here, too. In late summer, Brown Pelicans, Brandt's Cormorants, and, sometimes, hordes of Sooty Shearwaters fly by close to shore.

DIRECTIONS TO THE CLIFFS AT SHELL BEACH (MAP 15): From the south, exit U.S. 101 north of Pismo Beach at Mattie Rd./Shell Beach Rd. Turn left (west), then right (north) onto Shell Beach Rd. Follow Shell Beach Rd. for less than half a mile, until you see tennis courts on the left (west) side of the road. Park in the farthest west parking lot. Remember, this is private property, so please respect it. If you set up a spotting scope to look off the cliffs, nobody seems to mind. (*Note:* If you come from the north, you must exit U.S. 101 at Spyglass Rd., then turn left onto Shell Beach Rd. and drive approximately 2 miles to the tennis courts.)

DIRECTIONS TO MARGO DODD PARK (MAP 15): Another good place to observe Pigeon Guillemots (and Gray Whales *[Eschrichtius robustus]*, in season) is tiny Margo Dodd Park, on the bluffs a little upcoast from the tennis courts. From the tennis courts, drive north along Shell Beach Rd. to Cliff Ave. Turn left (west) on Cliff and follow it down to the bluff top. Park opposite the park with the gazebo. In May and June, look for the guillemots on the surrounding rock stacks, where they nest in crevices, or at the entrances to their burrows in the cliff face.

Montana de Oro State Park

Montana de Oro, a dramatic place of sand dunes and windswept bluffs, is as much a hiker's destination as a birder's. Bluff-top trails offer views of crashing breakers. Inland trails lead up moist canyons beside bubbling creeks. To see rocky shorebirds and migrating seabirds, walk on the trail that leads from the visitors parking lot out around Spooners Cove. You can walk for several miles up or down the coast from here. Watch for all three species of cormorant, nesting Pigeon Guillemots and Western Gulls in spring, and migrating seabirds such as loons, Brant, and scoters from mid-March to May. Montana de Oro State Park, Los Osos, CA 93402; (805) 528-0513; www.parks.ca.gov/default.asp.

DIRECTIONS (MAP 15): Exit U.S. 101 south of San Luis Obispo at Los Osos Valley Rd. Head west for 12 miles until the road turns left (south) and becomes Pecho Valley Rd. The park headquarters is at the end of Pecho Valley Rd.

Central San Luis Obispo County Coast (Map 16)

Sweet Springs Nature Preserve

The communities of Cuesta-by-the-Sea and Baywood Park boast several places from which to see the immense tidal mudflats for which Morro Bay is famous. Walk to the end of many of the streets, and you come to the edge of the bay. In winter at Sweet Springs, a nature preserve with a boardwalk and a viewing platform, all sorts of ducks can be seen when the tide is in. Look for the big flock of American Wigeons, which might have a Eurasian Wigeon (rare) with it. The Brant gather nearby, and American White Pelicans float in the distance. As the tide ebbs, the whole scenario changes. Hundreds of shorebirds, including American Avocets, Long-billed Curlews, and Marbled Godwits, feed on the exposed mudflats.

DIRECTIONS (MAP 16): From Los Osos Valley Rd., turn north onto Pine Ave. Follow Pine Ave. until it intersects with Ramona Ave. The preserve is across the street. Stay on the trails; poison oak *(Toxicodendron diversilobum)* thrives here.

Morro Bay Area

It is difficult to imagine a better birding getaway than the Morro Bay area in winter. Here are some suggestions, but there are many more places you can discover on your own.

DIRECTIONS (MAP 16): The town of Morro Bay is reached by taking California 1 west from U.S. 101 in San Luis Obispo.

Morro Bay State Park

State Park Marina Area. A walk out through the coastal sage scrub (several trails exist) past the boat basin brings you to the shore of the bay. At high tide, this is a good place to see shorebirds roosting. Resident White-crowned Sparrows hop around at the base of the bushes. A Peregrine Falcon may appear out of nowhere. A walk around to your left (east), brings you to an overlook from which to scan the shorebirds waiting out high tide. Out in the bay, Ospreys, terns, pelicans, and Brant enliven the scene.

DIRECTIONS (MAP 16): Take South Bay Blvd. south off California 1, bearing right (west) onto State Park Rd. Just beyond the state park entrance, pull over and park your car at the parking lot across the street from the boat basin.

Museum of Natural History. After a modernization project, the museum has reopened its doors, and it is very inviting, especially for young children. Although small, the museum commands a location perched high above Morro Bay. It has a good selection of checklists and natural history books in the gift store and offers many docent-led field trips in the Morro Bay area. Museum of Natural History, Morro Bay State Park, Morro Bay State Park Rd., Morro Bay, CA 93442; (805) 772-2694; www.morrobaymuseum.org.

DIRECTIONS (MAP 16): Proceed west on State Park Rd. past the boat marina to the museum, which is up the hill to the left.

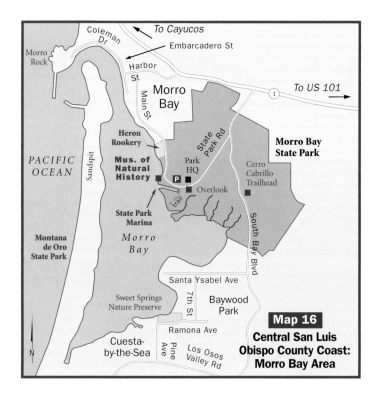

Figure 21. Peregrine Falcon.

Heron Rookery State Reserve. From February through June, a large colony of cormorants, herons, and egrets nests high in the tops of the eucalyptus trees. The colony is set off by a fence, but observers still have a good chance to study the domestic life of the birds.

DIRECTIONS (MAP 16): Continue west past the museum. A sign and parking pull-out mark the heron rookery on the left (west) side of State Park Rd.

Cerro Cabrillo Trailhead. Two trails, the Quarry Trail and the Live Oak Trail, trace through prime coastal sage scrub. The views of Morro Bay and Morro Rock are superb. In late winter and early spring, the shrubs and wildflowers are in bloom.

DIRECTIONS (MAP 16): The trailhead is on the east side of South Bay Blvd., approximately 1.4 miles north of the town of Baywood Park (midway between Baywood Park and the turn-off to State Park Rd.)

Morro Rock

This monolithic volcanic rock can be seen for miles around, situated across from the stacks of the power plant at the entrance to Morro Harbor. Morro Rock and the adjacent harbor waters make interesting birding. Numerous gulls and all three species of cormorant roost there. By far its most famous residents, the Peregrine Falcons that nest here (best seen in May and June) may be picked out by scanning the rocks above you. In the harbor waters, look for wintering loons and scoters, as well as American White Pelicans and Brant.

DIRECTIONS (MAP 16): To get to the base of Morro Rock from Morro Bay State Park, take State Park Rd. north, which becomes Main St. in the town of Morro Bay. Turn left (west) on Harbor St., and right (north) on Embarcadero St., which runs along the waterfront. Follow Embarcadero, which becomes Coleman Dr., to its terminus at the base of the rock.

Northern San Luis Obispo County Coast (Map 17)

Cayucos Pier

Walking out the pier in the beach town of Cayucos north of Morro Bay is peaceful and might turn up some interesting birds. In late summer and fall, Sooty Shearwaters come close inshore here. In winter, Black Turnstones, Surfbirds, and lots of gulls forage on the rocks. Scoters and grebes swim near the pier.

DIRECTIONS (MAP 17): At Cayucos, take the South Ocean Ave. exit off California 1. Drive north on South Ocean Ave. through the town. Turn left (south) on Cayucos Dr. to reach the base of the pier.

Estero Bluffs Trail

A gentle, grassy, coastal terrace is a good birding place at sunset on a winter day. A segment of land aquired by the Trust for Public Land and private donors to preserve the bluffs along the north shore of Estero Bay, this trail along the bluffs is part of the California Coastal

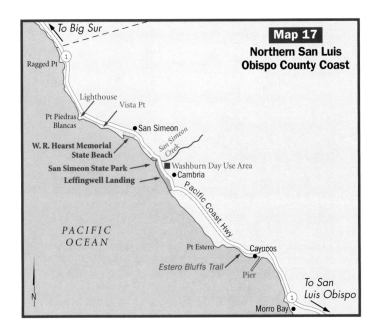

Trail. At low tide, Black Oystercatchers, Black Turnstones, Snowy Egrets, and Spotted Sandpipers investigate the rocks. In the fields, Northern Harriers, Red-tailed Hawks, Say's Phoebes, Savannah Sparrows, White-crowned Sparrows, and Western Meadowlarks can be found in fall and winter.

DIRECTIONS (MAP 17): Just north of Cayucos, watch for a point where the divided highway ends. Directly opposite San Geronimo Rd. on your right (north) is a parking pull-out on the coastal (south) side of California 1. Walk through an opening in the fence on a well-established trail that leads down to the coastal bluffs. Walk left (south) for a great view of Morro Rock in the distance.

San Simeon Coast

The birdfinding spots north of Cayucos vary considerably in character. Some are wide, open grasslands sloping to the edge of sandstone bluffs. Others are rocky seascapes. Still others are remnant wetlands. The grasslands boast a wide variety of hawks and falcons, especially in fall and winter. Look for Turkey Vultures, Northern Harriers, Red-tailed Hawks, Peregrine Falcons, and American Kestrels. Western Meadowlarks forage in the fields. Huge flocks of Brewer's and Red-winged Blackbirds appear where cattle graze. Along the rocky shores, resident Black Oystercatchers are joined by Black Turnstones, Surfbirds, and Wandering Tattlers. Each of the destinations below has elements of these types of birding. Each is found by turning off scenic California 1 to explore.

Cambria

North of the town of Cambria, the bluffs are part of San Simeon State Park. The best place for birding is at Leffingwell Landing.

DIRECTIONS (MAP 17): From California 1 at the traffic light in Cambria, take Windsor Blvd. west, then bear right (north) onto Moonstone Beach Dr. Near the end of Moonstone Beach Dr., turn left, following the sign to Leffingwell Landing.

Washburn Day Use Area, San Simeon Creek

This seasonal wetland has a boardwalk and a trail that meanders inland toward a Monterey pine *(Pinus radiata)* grove. In winter, ducks, herons, and egrets inhabit the wetland; grassland species forage in the meadows.

DIRECTIONS (MAP 17): From Leffingwell Landing, regain California 1 and head north (left turn). Take the next turn-off to your right, signed for San Simeon State Park, Washburn Day Use Area.

W. R. Hearst Memorial State Beach

San Simeon Bay is a stunning cove surrounded by a hamlet of Spanish-style buildings. Hearst's famous castle looks down on it from the far foothills.

Fortunately for nature-lovers and all who appreciate this splendid coast, the Hearst Corporation is currently negotiating a multimillion dollar conservation easement deal with the state of California to set aside some of its ranchland west of California 1 for preservation.

W. R. Hearst Memorial State Beach is a great picnic spot (restrooms), and there's a pier to walk out where loons, scoters, and gulls can be seen in fall and winter.

DIRECTIONS (MAP 17): Approximately 7 miles north of the Washburn Day Use Area, turn left (south) opposite the entrance to Hearst Castle State Historical Monument. This road is San Luis Obispo–San Simeon Rd. Follow it to the W. R. Hearst Memorial State Beach picnic grounds, which you can see from the frontage road.

Piedras Blancas Area

Vista Point, one mile south of Point Piedras Blancas, is interesting for birds, but the Northern Elephant Seals *(Mirounga angustirostris)* steal the show. From November through April, these enormous marine mammals lounge on the beach close by. The males display before their harems by engaging in fights with rivals and uttering snorts and roars. The scene is monitored by docents seven days a week in winter, and good interpretive signs make it a fascinating outing.

The lighthouse and its vicinity at Point Piedras Blancas are currently closed to the public; however, San Luis Obispo birders often set up their spotting scopes at the bluffs near the elephant seal colony to watch the spring migration of Pacific Loons, Brant, and scoters here in April.

DIRECTIONS (MAP 17): Approximately 4 miles north of W. R. Hearst Memorial State Beach, turn left (west) where a sign says Vista Point and shows the "brown binoculars" wildlife viewing symbol.

GOING BY BOAT

Short trips offshore to visit the Channel Islands (see map 1, page 5) or along the coast for whale-watching can be fun. Occasionally, these boat trips produce good sightings of pelagic bird species. In addition to Sooty Shearwaters (spring through fall) and Black-vented Shearwaters (late fall and winter), you may see Pink-footed Shearwaters *(Puffinus creatopus)* and Black Storm-Petrels *(Oceanodroma melania)* (spring through summer), Pomarine and Parasitic Jaegers *(Stercorarius pomarinus* and *S. parasiticus)* (spring or fall), Cassin's Auklet *(Ptychoramphus aleuticus)* (year-round), and Rhinoceros Auklet *(Cerorhinca monocerata)* (fall through early spring).

Catalina Island Ferry

To take a ferry to Santa Catalina Island, consider one of the three boat services that depart from San Pedro, Long Beach, Newport Beach, and Dana Point. Reservations are recommended. A good Web site is www.ecatalina.com/transportation.asp. Catalina Express operates from San Pedro, Long Beach Downtown Landing, Long Beach Queen Mary, and Dana Point; (310) 519-1212 or (800) 618-5533. Catalina Explorer Co. departs from Dana Point; (800) 432-6276. Catalina Passenger Service operates the Catalina Flyer out of Newport Beach; (949) 673-5245.

Channel Islands National Park

To visit the islands that make up Channel Islands National Park — Anacapa, Santa Cruz, Santa Rosa, San Miguel, and Santa Barbara — board a boat in Ventura or Channel Islands Harbors. In spring, Brown Pelicans (not on Santa Cruz Island), Double-crested, Brandt's, and Pelagic Cormorants, Black Oystercatchers, and Pigeon Guillemots nest on Anacapa and Santa Cruz Islands.

Island Packers, a private concessionaire located next door to the national park headquarters, has a full schedule of boat trips to the park. Inquire at Island Packers, 1867 Spinnaker Dr., Ventura, CA 93001; call (805) 642-7688 for a recorded schedule and fare information, and (805) 642-1393 for reservations. The shortest trip is the one to East Anacapa Island. Visiting the other islands takes a little longer.

Whale-watching

Whale-watching trips usually last a half-day or less and are sometimes a good way to become initiated in pelagic birding. Gray Whales make their annual migration from Alaska to Baja California (where they calve) in late December and January. On the whales' return trip, in March and April, they once again pass by southern California's shores. In summer, Blue and Humpback Whales can be seen in the Santa Barbara Channel. The whole whale-watching experience is fascinating, and lots of marinas and landings have whale-watching excursions. There are numerous whale-watching opportunities from boats leaving harbors along the coasts of San Diego County (Shelter

Island, Mission Bay), Orange County (Dana Point), and Los Angeles County (Redondo Beach, Long Beach). Find your own, or pick from the following: in Santa Barbara, The Condor Express at Sea Landing in Santa Barbara Harbor (805) 882-0088 or (800) 77WHALE, and Island Packers (see above); in Morro Bay, Virg's Landing (805) 772-1222, and Bob's Sportfishing (805) 772-3340.

SEASONAL OCCURRENCE
BAR GRAPHS

Abundance	Breeding Status					
■ common	B = breeder					
■ fairly common	B* = irregular, scarce, or					
— uncommon	very local breeder					
- - - rare		MARCH	APRIL	MAY	JUNE	JULY

GEESE AND DUCKS (ANATIDAE)

☐ Canada Goose (*Branta canadensis*), B*

☐ Brant (*Branta bernicla*)

☐ Gadwall (*Anas strepera*), B

☐ American Wigeon (*Anas americana*)

☐ Mallard (*Anas platyrhynchos*), B

☐ Blue-winged Teal (*Anas discors*), B*

☐ Cinnamon Teal (*Anas cyanoptera*), B

☐ Northern Shoveler (*Anas clypeata*)

☐ Northern Pintail (*Anas acuta*)

☐ Green-winged Teal (*Anas crecca*)

☐ Canvasback (*Aythya valisineria*)

☐ Redhead (*Aythya americana*), B*

☐ Ring-necked Duck (*Aythya collaris*)

☐ Lesser Scaup (*Aythya affinis*)

☐ Surf Scoter (*Melanitta perspicillata*)

☐ Bufflehead (*Bucephala albeola*)

☐ Red-breasted Merganser (*Mergus serrator*)

☐ Ruddy Duck (*Oxyura jamaicensis*), B

LOONS (GAVIIDAE)

☐ Red-throated Loon (*Gavia stellata*)

☐ Pacific Loon (*Gavia pacifica*)

☐ Common Loon (*Gavia immer*)

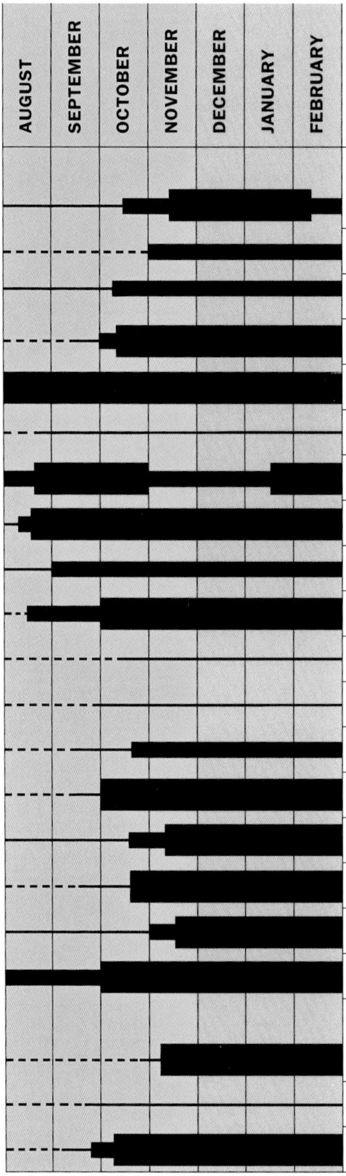
AUGUST
SEPTEMBER
OCTOBER
NOVEMBER
DECEMBER
JANUARY
FEBRUARY

NOTES

Abundance

■ common
■ fairly common
— uncommon
- - - rare

Breeding Status
B = breeder
B* = irregular, scarce, or
       very local breeder

MARCH | APRIL | MAY | JUNE | JULY

GREBES (PODICIPEDIDAE)

☐ Pied-billed Grebe *(Podilymbus podiceps)*, B

☐ Horned Grebe *(Podiceps auritus)*

☐ Eared Grebe *(Podiceps nigricollis)*, B*

☐ Western Grebe *(Aechmophorus occidentalis)*

☐ Clark's Grebe *(Aechmophorus clarkii)*

SHEARWATERS (PROCELLARIIDAE)

☐ Sooty Shearwater *(Puffinus griseus)*

☐ Black-vented Shearwater *(Puffinus opisthomelas)*

PELICANS (PELECANIDAE)

☐ American White Pelican *(Pelecanus erythrorhynchos)*

☐ Brown Pelican *(Pelecanus occidentalis)*, B*

CORMORANTS (PHALACROCORACIDAE)

☐ Brandt's Cormorant *(Phalacrocorax penicillatus)*, B

☐ Double-crested Cormorant *(Phalacrocorax auritus)*, B

☐ Pelagic Cormorant *(Phalacrocorax pelagicus)*, B

HERONS, EGRETS, AND BITTERNS (ARDEIDAE)

☐ Great Blue Heron *(Ardea herodias)*, B

☐ Great Egret *(Ardea alba)*, B*

☐ Snowy Egret *(Egretta thula)*, B*

☐ Green Heron *(Butorides virescens)*, B

☐ Black-crowned Night-Heron *(Nycticorax nycticorax)*, B

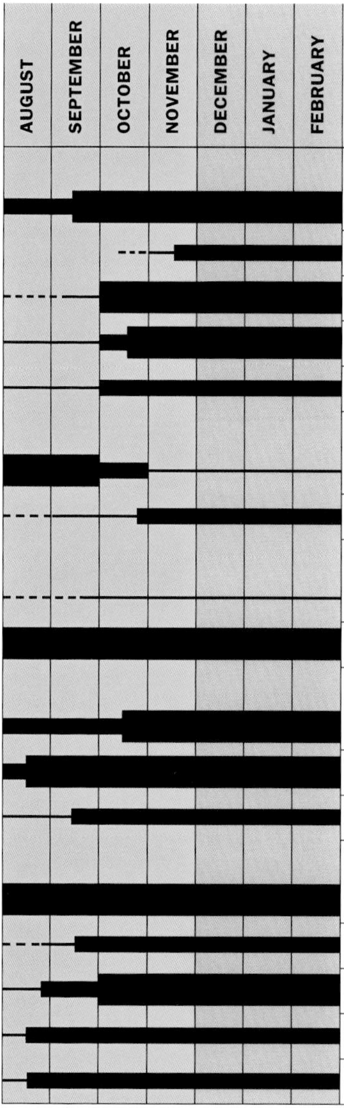
AUGUST
SEPTEMBER
OCTOBER
NOVEMBER
DECEMBER
JANUARY
FEBRUARY

NOTES

Abundance

■ common
■ fairly common
— uncommon
- - - rare

Breeding Status

B = breeder
B* = irregular, scarce, or
      very local breeder

	MARCH	APRIL	MAY	JUNE	JULY

IBISES (THRESKIORNITHIDAE)

☐ White-faced Ibis *(Plegadis chihi)*

NEW WORLD VULTURES (CATHARTIDAE)

☐ Turkey Vulture *(Cathartes aura)*

HAWKS (ACCIPITRIDAE)

☐ Osprey *(Pandion haliaetus)*

☐ White-tailed Kite *(Elanus leucurus)*, B

☐ Northern Harrier *(Circus cyaneus)*, B*

☐ Red-shouldered Hawk *(Buteo lineatus)*, B

☐ Red-tailed Hawk *(Buteo jamaicensis)*, B

FALCONS (FALCONIDAE)

☐ American Kestrel *(Falco sparverius)*, B

☐ Merlin *(Falco columbarius)*

☐ Peregrine Falcon *(Falco peregrinus)*, B*

RAILS, GALLINULES, AND COOTS (RALLIDAE)

☐ Clapper Rail *(Rallus longirostris)*, B*

☐ Virginia Rail *(Rallus limicola)*, B*

☐ Sora *(Porzana carolina)*, B*

☐ Common Moorhen *(Gallinula chloropus)*, B*

☐ American Coot *(Fulica americana)*, B

PLOVERS (CHARADRIIDAE)

☐ Black-bellied Plover *(Pluvialis squatarola)*

☐ Snowy Plover *(Charadrius alexandrinus)*, B*

AUGUST	SEPTEMBER	OCTOBER	NOVEMBER	DECEMBER	JANUARY	FEBRUARY	NOTES

Abundance
■ common
■ fairly common
— uncommon
- - - rare

Breeding Status
B = breeder
B* = irregular, scarce, or
    very local breeder

MARCH APRIL MAY JUNE JULY

☐ Semipalmated Plover *(Charadrius semipalmatus)*

☐ Killdeer *(Charadrius vociferus)*, B

OYSTERCATCHERS (HAEMATOPODIDAE)

☐ Black Oystercatcher *(Haematopus bachmani)*, B

STILTS AND AVOCETS (RECURVIROSTRIDAE)

☐ Black-necked Stilt *(Himantopus mexicanus)*, B

☐ American Avocet *(Recurvirostra americana)*, B

SANDPIPERS (SCOLOPACIDAE)

☐ Greater Yellowlegs *(Tringa melanoleuca)*

☐ Lesser Yellowlegs *(Tringa flavipes)*

☐ Willet *(Catoptrophorus semipalmatus)*

☐ Wandering Tattler *(Heteroscelus incanus)*

☐ Spotted Sandpiper *(Actitis macularia)*, B*

☐ Whimbrel *(Numenius phaeopus)*

☐ Long-billed Curlew *(Numenius americanus)*

☐ Marbled Godwit *(Limosa fedoa)*

☐ Ruddy Turnstone *(Arenaria interpres)*

☐ Black Turnstone *(Arenaria melanocephala)*

☐ Surfbird *(Aphriza virgata)*

☐ Red Knot *(Calidris canutus)*

☐ Sanderling *(Calidris alba)*

☐ Western Sandpiper *(Calidris mauri)*

AUGUST	SEPTEMBER	OCTOBER	NOVEMBER	DECEMBER	JANUARY	FEBRUARY	NOTES

Abundance

■ common
■ fairly common
— uncommon
- - - rare

Breeding Status
B = breeder
B* = irregular, scarce, or very local breeder

	MARCH	APRIL	MAY	JUNE	JULY

☐ Least Sandpiper *(Calidris minutilla)*

☐ Dunlin *(Calidris alpina)*

☐ Short-billed Dowitcher *(Limnodromus griseus)*

☐ Long-billed Dowitcher *(Limnodromus scolopaceus)*

☐ Wilson's Snipe *(Gallinago delicata)*

☐ Wilson's Phalarope *(Phalaropus tricolor)*

☐ Red-necked Phalarope *(Phalaropus lobatus)*

GULLS, TERNS, AND SKIMMERS (LARIDAE)

☐ Bonaparte's Gull *(Larus philadelphia)*

☐ Heermann's Gull *(Larus heermanni)*

☐ Mew Gull *(Larus canus)*

☐ Ring-billed Gull *(Larus delawarensis)*

☐ California Gull *(Larus californicus)*

☐ Herring Gull *(Larus argentatus)*

☐ Western Gull *(Larus occidentalis)*, B

☐ Glaucous-winged Gull *(Larus glaucescens)*

☐ Caspian Tern *(Sterna caspia)*, B

☐ Royal Tern *(Sterna maxima)*, B*

☐ Elegant Tern *(Sterna elegans)*, B

☐ Forster's Tern *(Sterna forsteri)*, B

☐ Least Tern *(Sterna antillarum)*, B*

☐ Black Skimmer *(Rynchops niger)*, B

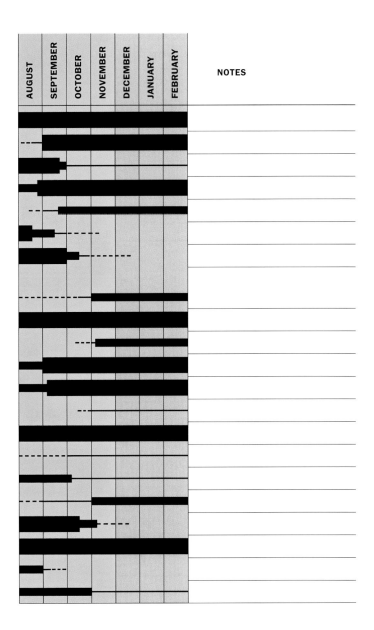

Abundance

■ common
■ fairly common
— uncommon
- - - rare

Breeding Status

B = breeder
B* = irregular, scarce, or
     very local breeder

	MARCH	APRIL	MAY	JUNE	JULY

AUKS (ALCIDAE)

☐ Pigeon Guillemot *(Cepphus columba)*, B

BARN OWLS (TYTONIDAE)

☐ Barn Owl *(Tyto alba)*, B

KINGFISHERS (ALCEDINIDAE)

☐ Belted Kingfisher *(Ceryle alcyon)*, B*

TYRANT FLYCATCHERS (TYRANNIDAE)

☐ Black Phoebe *(Sayornis nigricans)*, B

☐ Say's Phoebe *(Sayornis saya)*

CROWS AND RAVENS (CORVIDAE)

☐ American Crow *(Corvus brachyrhynchos)*, B

☐ Common Raven *(Corvus corax)*, B

SWALLOWS AND MARTINS (HIRUNDINIDAE)

☐ Tree Swallow *(Tachycineta bicolor)*, B

☐ Cliff Swallow *(Petrochelidon pyrrhonota)*, B

☐ Barn Swallow *(Hirundo rustica)*, B

WRENS (TROGLODYTIDAE)

☐ Marsh Wren *(Cistothorus palustris)*, B

STARLINGS AND MYNAS (STURNIDAE)

☐ European Starling *(Sturnus vulgaris)*, B

WAGTAILS AND PIPITS (MOTACILLIDAE)

☐ American Pipit *(Anthus rubescens)*

WOOD-WARBLERS (PARULIDAE)

☐ Common Yellowthroat *(Geothlypis trichas)*, B

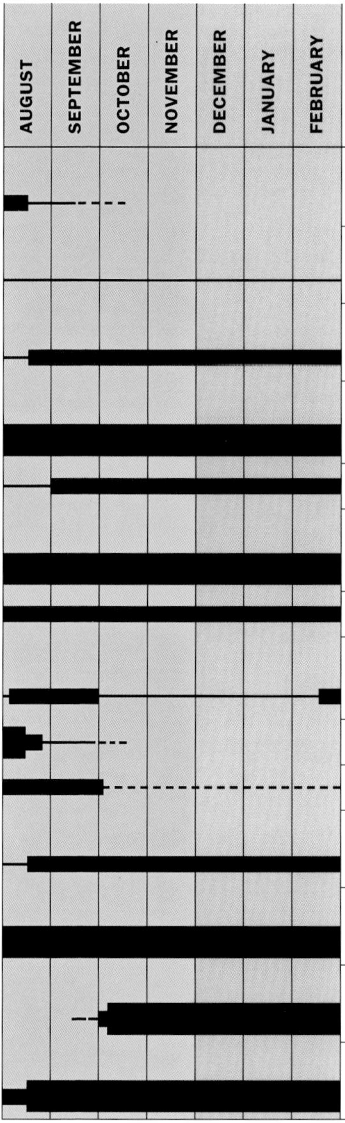
AUGUST
SEPTEMBER
OCTOBER
NOVEMBER
DECEMBER
JANUARY
FEBRUARY

NOTES

Abundance	Breeding Status						
■ common	B = breeder						
■ fairly common	B* = irregular, scarce, or						
— uncommon	very local breeder						
– – – rare		MARCH	APRIL	MAY	JUNE	JULY	

SPARROWS (EMBERIZIDAE)

☐ Spotted Towhee *(Pipilo maculatus)*, B

☐ California Towhee *(Pipilo crissalis)*, B

☐ Savannah Sparrow *(Passerculus sandwichensis)*

☐ Belding's Savannah Sparrow
  (Passerculus sandwichensis beldingi), B*

☐ Song Sparrow *(Melospiza melodia)*, B

☐ White-crowned Sparrow *(Zonotrichia leucophrys)*, B*

☐ Golden-crowned Sparrow *(Zonotrichia atricapilla)*

BLACKBIRDS AND ORIOLES (ICTERIDAE)

☐ Red-winged Blackbird *(Agelaius phoeniceus)*, B

☐ Western Meadowlark *(Sturnella neglecta)*, B

☐ Brewer's Blackbird *(Euphagus cyanocephalus)*, B

☐ Hooded Oriole *(Icterus cucullatus)*, B

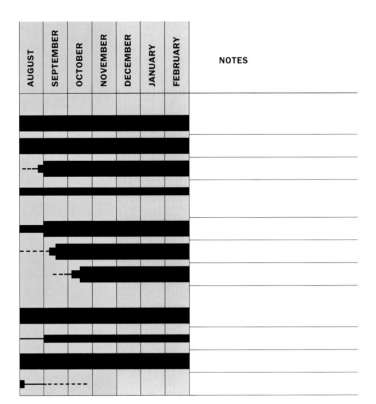

AUGUST	SEPTEMBER	OCTOBER	NOVEMBER	DECEMBER	JANUARY	FEBRUARY	NOTES

SHOREBIRD SIZE CHART

This chart and table help to organize by size the shorebirds found along the southern California coast. Shorebirds are those species that live near water along the coast, such as plovers, oystercatchers, stilts, avocets, and sandpipers. The size of the species often dictates its habits and gives clues to its identity.

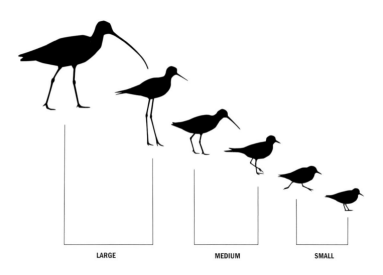

LARGE   MEDIUM   SMALL

Shorebird silhouettes in order of descending size (left to right): Long-billed Curlew, Black-necked Stilt, Long-billed Dowitcher, Killdeer, Sanderling, and Least Sandpiper.

Body Length of Shorebirds of the
Southern California Coast (in Inches)

LARGE SHOREBIRDS	
Long-billed Curlew	23
American Avocet	18
Marbled Godwit	18
Whimbrel	17.5
Black Oystercatcher	17.5
Willet	15
Black-necked Stilt	14
Greater Yellowlegs	14

MEDIUM SHOREBIRDS	
Long-billed Dowitcher	11.5
Black-bellied Plover	11.5
Short-billed Dowitcher	11
Wandering Tattler	11
Wilson's Snipe	10.5
Red Knot	10.5
Killdeer	10.5
Lesser Yellowlegs	10.5
Surfbird	10
Ruddy Turnstone	9.5
Black Turnstone	9.25
Wilson's Phalarope	9.25

SMALL SHOREBIRDS	
Dunlin	8.5
Sanderling	8
Red-necked Phalarope	7.75
Spotted Sandpiper	7.5
Semipalmated Plover	7.25
Western Sandpiper	6.5
Snowy Plover	6.25
Least Sandpiper	6

GULL IDENTIFICATION

The study of gulls is challenging. Gulls are confusing to identify because they go through a series of molts as they mature. Each species of gull has its own rate of maturation: the larger the gull the longer it takes to make the progression from juvenal plumage to that of adulthood. For example, a large gull such as the Western Gull takes four years to achieve adult plumage. The small Bonaparte's Gull takes only two years.

If you look at a group of gulls gathered on the beach in winter, you notice that some individuals have a mottled, brownish appearance all over; others have a mottled, brownish appearance with a grayish back and a white head; still others have no brownish feathers, but gray backs and wings and white heads. In addition, they have various bill and leg colors.

Since gulls show very little sexual dimorphism (i.e., the males and females appear alike) the plumage differences you see in the group of gulls on the beach are tied to the age of the birds. Furthermore, the gathering could be an assortment of species of gulls.

The difficult aspect of gull identification is twofold: not only might there be a variety of gull species in that group on the beach, but each gull species looks a little different as it progresses from its first year toward adulthood. Most immature gulls have a mottled, brownish appearance in their first year, then progress through various molts to a gray-backed, white-headed adult plumage.

True gull aficionados will wish to study gull plumage changes in depth. For beginners, however, the intricacies of gull identification can be overwhelming. As a starting point, learn the adult plumages of the common gulls. Once that has been accomplished, a knowledge of first-year (known as " first-winter") plumages is extremely helpful. Knowing the first-winter and adult plumages is a foundation for gull study. From there, you can build a mental picture of the progression the species undergoes to bridge the gap between the first-winter plumage and that of the adult.

Take the example of a California Gull *(Larus californicus)*, a four-year gull. In its first winter, a young California Gull has a brownish, mottled look; its bill is pink with a black tip; its legs are pink, too. By its second winter, the California Gull has acquired a grayish back, but the rest of its plumage is still mottled brown; its bill is now blue gray with a black spot toward the tip, and its legs are now blue gray. By its third winter, this California Gull is beginning to resemble an adult. It lacks the brownish feathers of the earlier years and shows a gray back and wings, whitish head and belly, black wing tips, yellow legs, and a yellow bill with a black spot toward the tip. The fourth-year bird is the classic California Gull: gray back and wings, black wing tips with white spots, yellow bill with a black and red spot near the tip. (Several species of gulls have streaking on their heads in adult nonbreeding plumage, but this is not important to basic gull identification.)

In the last 25 years, good field guides have begun to show detailed gull plumage progressions—an incentive for birders to become more comfortable with gull identification. Because the gull flocks along the southern California coast in late fall and winter are composed of adults and immatures of a variety of species, it is an excellent place to start taking an interest in gulls.

TABLE 4. Relative Sizes and Plumage Progressions of Common Southern California Gulls

Large	4 years to adult plumage	Glaucous-winged Gull
		Western Gull
		Herring Gull
		California Gull
		Heermann's Gull
Medium	3 years to adult plumage	Ring-billed Gull
		Mew Gull
Small	2 years to adult plumage	Bonaparte's Gull

SUGGESTED READING

Beginning Birding

Dunne, Peter. 2003. *Pete Dunne on bird watching: The how-to, when-to, and where-to of birding.* Boston: Houghton Mifflin.

Sibley, David Allen. 2002. *Birding basics.* New York: Alfred A. Knopf.

Field Guides

Griggs, Jack L. 1997. *All the birds of North America.* New York: Harper-Collins.

Kaufman, Kenn. 2000. *Birds of North America.* Kaufman Focus Guides. New York: Houghton Mifflin.

National Geographic Society. 2002. *Field guide to the birds of North America.* 4th ed. Washington, D.C.: National Geographic Society.

Peterson, Roger Tory. 1990. *Western birds.* Boston: Houghton Mifflin.

Sibley, David Allen. 2000. *The Sibley guide to birds.* New York: Alfred A. Knopf.

Sibley, David Allen. 2003. *The Sibley field guide to birds of western North America.* New York: Alfred A. Knopf.

Stokes, Donald, and Lillian Stokes. 1996. *Stokes field guide to birds: Western region.* Boston: Little Brown.

Regional Books

Bakker, Elna. 1984. *An island called California.* 2nd ed. Berkeley and Los Angeles: University of California Press.

Clarke, Herbert. 1984. *An introduction to Southern California birds.* Missoula, Mont.: Mountain Press Publishing.

Cogswell, Howard L. 1977. *Water birds of California.* Berkeley and Los Angeles: University of California Press.

Hamilton, Robert A., and Douglas R. Willick. 1996. *The birds of Orange County, California: Status and distribution.* Irvine, Calif.: Sea and Sage Audubon Society.

Lehman, Paul E. 1994. *The birds of Santa Barbara County, California.* Lawrence, Kans.: Allen Press/Vertebrate Museum, University of California, Santa Barbara.

Lehman, Paul E., ed. 2001. *A birder's guide to metropolitan areas of North America.* Colorado Springs, Colo.: American Birding Association.

McCausland, Bill. 2000. *Birding San Diego County.* San Diego: San Diego Audubon Society.

Paulson, Dennis. 1993. *Shorebirds of the Pacific Northwest.* Seattle: University of Washington Press.

Schoenherr, Allan A. 1992. *A natural history of California.* Berkeley and Los Angeles: University of California Press.

Schram, Brad. 1998. *A birder's guide to southern California.* Colorado Springs, Colo.: American Birding Association.

Small, Arnold. 1994. *California birds: Their status and distribution.* Vista, Calif.: Ibis Publishing.

Stallcup, Rich. 1990. *Ocean birds of the nearshore Pacific.* Stinson Beach, Calif.: Point Reyes Bird Observatory.

Unitt, Philip. 1984. *The birds of San Diego County.* San Diego: San Diego Society of Natural History.

Zimmer, Kevin J. 2000. *Birding in the American West: A handbook.* Ithaca, N.Y.: Cornell University Press.

Regional Checklists

American Birding Association. 2002. *ABA Checklist,* 6th ed. Colorado Springs, Colo.: American Birding Association.

Edell, Tom. 2002. *The birds of San Luis Obispo County, California.* Morro Bay, Calif.: Morro Coast Audubon Society.

Garrett, Kimball L., and Mike San Miguel. 2000. *Field list of the birds of Los Angeles County.* Los Angeles: Los Angeles Audubon Society.

Lehman, Paul, and Joan Lentz. 1993. *Birds of Santa Barbara County.* Goleta, Calif.: Santa Barbara Audubon Society.

McCaskie, Guy. 2000. *Field checklist of the birds of San Diego County.* San Diego: County of San Diego Department of Parks and Recreation.

National Audubon Society. N.d. *Central coast birding trail, Santa Barbara County.* N.p.: National Audubon Society, Central Coast Chapters.

National Audubon Society. N.d. *Central coast birding trail, Ventura County.* N.p.: National Audubon Society, Central Coast Chapters.

Sea and Sage Audubon Society. 1991. *The birds of Orange County.* Irvine, Calif.: Sea and Sage Press.

Smith, Reed V., Linda O'Neill and Steve Tucker. N.d. "Checklist of the Birds of Ventura County California." Ventura, Cal.: Ventura Audobon Society.

General Reference

California Coastal Commission. 2003. *California coastal resource guide.* 6th ed. Berkeley and Los Angeles: University of California Press.

California Department of Fish and Game. 2005. State and federally endangered and threatened animals of California. California Natural Diversity Database. www.dfg.ca.gov/.

Dailey, Murray D., Donald J. Reish, and Jack W. Anderson, eds. 1993. *Ecology of the Southern California Bight.* Berkeley and Los Angeles: University of California Press.

Ehrlich, Paul R., David S. Dobkin, and Darryl Wheye. 1988. *The birder's handbook.* New York: Simon and Schuster.

Gill, Frank. 1994. *Ornithology.* 2nd ed. New York: W. H. Freeman.

Harrison, Peter. 1985. *Seabirds: An identification guide.* Boston: Houghton Mifflin.

Heinrich, Bernd. 2004. *The geese of Beaver Bog.* New York: HarperCollins.

Kaufman, Kenn. 1996. *Lives of North American birds.* Boston: Houghton Mifflin.

Poole, Alan, and Frank Gill, eds. 1992–2002. *The birds of North America: Life histories for the twenty-first century.* 18 vols. Philadelphia: The American Ornithologists' Union and The Academy of Natural Sciences. http://bna.birds.cornell.edu/BNA/

Sibley, David Allen. 2001. *The Sibley guide to bird life and behavior.* New York: Alfred A. Knopf.

Terres, John K. 1980. *The Audubon Society encyclopedia of North American birds.* New York: Alfred A. Knopf.

Migration

Matthiessen, Peter. 1973. *The wind birds.* Houghton Mifflin.

Tennant, Alan. 2004. *On the wing: To the ends of the earth with the Peregrine Falcon.* New York: Alfred A. Knopf.

Weidensaul, Scott. 1999. *Living on the wind: Across the hemisphere with migratory birds.* New York: North Point Press.

Older Titles Worth Hunting For

Bent, Arthur Cleveland, ed. 1919–1968. *Life histories of North American birds.* 26 vols. Repr., New York: Dover. Many volumes still in print.

Davis, John, and Alan Baldridge. 1980. *The bird year: A book for birders.* Pacific Grove, Calif.: The Boxwood Press.

Dawson, William Leon. 1923. *The birds of California.* 3 vols. San Diego: South Moulton.

Garrett, Kimball, and Jon Dunn. 1981. *Birds of southern California: Status and distribution.* Los Angeles: Los Angeles Audubon Society.

Hoffmann, Ralph. 1927. *Birds of the Pacific states.* Boston: Houghton Mifflin.

Periodicals

Audubon. National Audubon Society, 700 Broadway, New York, NY 10003.

Birding. American Birding Association, P.O. Box 6599, Colorado Springs, CO 80934.

Bird Watcher's Digest. P. O. Box 110, Marietta, OH 45750.

Living Bird. Cornell University Laboratory of Ornithology, 159 Sapsucker Woods Rd., Ithaca, NY 14850.

Natural History. American Museum of Natural History, Central Park West and 79th St., New York, NY 10024.

North American Birds. American Birding Association, P.O. Box 6599, Colorado Springs, CO 80934.

Professional Journals

The Auk. American Ornithologists' Union, 300 West Chestnut St., Ephrata, PA 17522.

The Condor. Cooper Ornithological Society, High Desert Ecological Research Institute, 15 SW Colorado, Ste. 300, Bend, OR 97702.

The Wilson Bulletin. Wilson Ornithological Society, University of Michigan Museum of Zoology, Ann Arbor, MI 48109.

INDEX OF BIRDS

GENERAL INDEX

ABOUT THE AUTHOR

Joan Easton Lentz is a research associate with the Santa Barbara Museum of Natural History. She also teaches bird classes for the Santa Barbara Community College Continuing Education Division and is the author of *Great Birding Trips of the West* (1989) and *Birdwatching: A Guide for Beginners* (1985).

ABOUT THE PHOTOGRAPHER

Don DesJardin is a longtime resident of Ventura, California, where he was introduced to the hobby of birding and bird photography in 1989. Well published both in print and on the Internet, he enjoys not only the challenges and rewards of bird photography but also sharing his photos with others.

ABOUT THE ILLUSTRATOR

Peter Gaede has been a full-time natural science illustrator since graduating from the Scientific Illustration Program at the University of California, Santa Cruz, in 2000. His illustrations have appeared in a variety of natural history books and magazines, scientific journals, museums, and nature centers. He currently resides in Carpinteria, California.

Series Design:	Barbara Jellow
Design Enhancements:	Beth Hansen
Design Development:	Jane Tenenbaum
Cartographer:	Hayden Foell
Composition:	Jane Tenenbaum
Text:	9/10.5 Minion
Display:	Franklin Gothic Book and Demi
Printer and binder:	Everbest Printing Company

CALIFORNIA NATURAL HISTORY GUIDES

Field Guides

Sharks, Rays, and Chimaeras of California
David A. Ebert. Illustrations by Mathew D. Squillante
0-520-22265-2 cloth, 0-520-23484-7 paper

Mammals of California
Revised Edition. E.W. Jameson, Jr., and Hans J. Peeters
0-520-23581-9 cloth, 0-520-23582-7 paper

Dragonflies and Damselflies of California
Timothy D. Manolis
0-520-23566-5 cloth, 0-520-23567-3 paper

Freshwater Fishes of California
Revised Edition. Samuel M. McGinnis.
Illustrations by Doris Alcorn
0-520-23728-5 cloth, 0-520-23727-7 paper

Trees and Shrubs of California
John D. Stuart and John O. Sawyer
0-520-22109-5 cloth, 0-520-22110-9 paper

Pests of the Native California Conifers
David L. Wood, Thomas W. Koerber, Robert F. Scharpf, and
Andrew J. Storer
0-520-23327-1 cloth, 0-520-23329-8 paper

Introductory Guides

Introduction to Water in California
David Carle
0-520-23580-0 cloth, 0-520-24086-3 paper

Introduction to California Beetles
Arthur V. Evans and James N. Hogue
0-520-24034-0 cloth, 0-520-24035-9 paper

Weather of the San Francisco Bay Region
Second Edition. Harold Gilliam
0-520-22989-4 cloth, 0-520-22990-8 paper

Introduction to Trees of the San Francisco Bay Region
Glenn Keator
0-520-23005-1 cloth, 0-520-23007-8 paper

Introduction to Shore Wildflowers of California, Oregon, and Washington
Revised Edition. Philip A. Munz. Edited by Dianne Lake and Phyllis M. Faber. Introduction by Robert Orcduff
0-520-23638-6 cloth, 0-520-23639-4 paper

Introduction to California Mountain Wildflowers
Revised Edition. Philip A. Munz. Edited by Dianne Lake and Phyllis M. Faber. Introduction by Robert Orcduff
0-520-23635-1 cloth, 0-520-23637-8 paper

Introduction to California Desert Wildflowers
Revised Edition. Philip A. Munz. Edited by Diane L. Renshaw and Phyllis M. Faber. Introduction by Robert Orcduff
0-520-23631-9 cloth, 0-520-23632-7 paper

Introduction to California Spring Wildflowers of the Foothills, Valleys, and Coast
Revised Edition. Philip A. Munz. Edited by Dianne Lake and Phyllis M. Faber. Introduction by Robert Orcduff
0-520-23633-5 cloth, 0-520-23634-3 paper

Introduction to California Plant Life
Revised Edition. Robert Orcduff, Phyllis M. Faber, and Todd Keeler-Wolf
0-520-23702-1 cloth, 0-520-23704-8 paper

Introduction to Horned Lizards of North America
Wade C. Sherbrooke
0-520-22825-1 cloth, 0-520-22827-8 paper

Regional Guides

Sierra Nevada Natural History
Revised Edition. Tracy I. Storer, Robert L. Usinger, and David Lukas
0-520-23277-1 cloth, 0-520-24096-0 paper